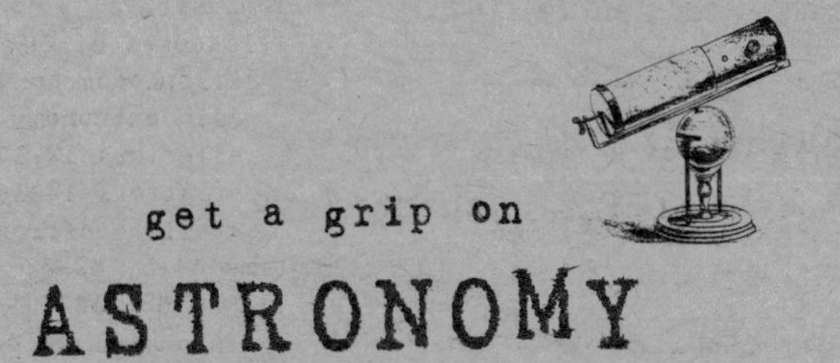

get a grip on

ASTRONOMY

Get a Grip on
ASTRONOMY

ROBIN KERROD

Time-Life Books is a division of Time Life Inc.

TIME LIFE INC.
President and CEO: George Artandi

TIME-LIFE CUSTOM PUBLISHING

Vice President and Publisher	Terry Newell
Vice President of Sales and Marketing	Neil Levin
Project Manager	Jennie Halfant
Director of Creative Services	Laura Ciccone McNeill
Director of Acquisitions	Jennifer Pearce
Director of Special Markets	Liz Ziehl

This book was conceived, designed, and produced by The Ivy Press Limited,
2-3 St. Andrews Place, Lewes, East Sussex BN7 1UP, England

Art Director	Peter Bridgewater
Editorial Director	Sophie Collins
Designer	Angela English
Commissioning Editor	Andrew Kirk
Picture Research	Vanessa Fletcher
Illustrations	Andrew Kulman

Reproduction and printing in Hong Kong by
Hong Kong Graphic and Printing Ltd.

LIBRARY OF CONGRESS CATALOGING-IN-PUBLICATION DATA
Kerrod, Robin.
Astronomy/Robin Kerrod.
p. cm. -- (Get a grip on--)
Includes index.
ISBN 0-7370-0047-3 (softcover : alk. paper)
1. Astronomy--Popular works. I. Title. II. Series.
QB44.2.K44 1999
520--dc21
98-53078
CIP

CONTENTS

INTRODUCTION

ASTRONOMY IS LOOKING UP

"No, I'm sure that's Orion over there!"

★ On a clear night, treat yourself to the greatest free show on Earth. Look up at the sky, and you'll see the dazzling spectacle of thousands of stars sparkling like jewels on black velvet. And you may find it hard to suppress that delightful ditty "Twinkle, twinkle, little star, How I wonder what you are!"

BUILDING FOUNDATIONS

★ Thousands of years ago, when our ancestors looked up at the night sky—THE HEAVENS—they began to wonder about what they saw there. Being of a superstitious frame of mind, they found the heavens awesome. Strange things happening there (the appearance of a comet, for example) seemed to them to be portents of disaster—death, destruction, pestilence, drought, or flood.

★ *The learned priests who guided early civilizations started to study the heavens seriously, to look for omens prophetic of good or evil, which is how* ASTROLOGY *began. Although, from the scientific point of view, they studied the heavens for the wrong reasons, their observations laid the foundations of* ASTRONOMY.

KEY WORDS

THE HEAVENS: the night sky

HEAVENLY BODIES: the objects seen in the night sky, such as the moon and stars

TELESCOPE: an optical instrument that gathers much more light than the human eye and magnifies distant objects (literally, the word means a device that "sees far")

UNIVERSE: everything that exists—space and all the heavenly bodies that occupy it

WITH THE NAKED EYE: using your eyes alone, unaided by a telescope or binoculars

ASTRONOMERS DO IT ALL NIGHT

★ Some five millennia on, present-day astronomers can justifiably claim that theirs is the oldest science. There's little old-fashioned, however, about modern astronomy. True, astronomers still spend their nights stargazing, but many of them have powerful TELESCOPES that can peer to the edges of the UNIVERSE. Others use RADIO TELESCOPES and SPACE SATELLITES to view objects in the universe at wavelengths human eyes can't detect; or despatch PROBES billions of miles to observe or land on remote planets.

observing the stars is so fascinating, some astronomers do it all night

★ *With these new tools, astronomers are making new discoveries all the time and finding that the deeper they probe into the universe, the more mysterious it becomes. They seem continually to be finding answers for which they don't even have questions.*

STARRY-EYED

There are many millions of stars in the sky—but we can only see a few thousand of them with the naked eye. In the whole heavens, fewer than 6,000 stars are bright enough to be visible; and at any one time, less than 2,500 stars are visible above the horizon.

What is astronomy?

ASTRONOMY *is the scientific study of the heavens and all that is therein. It is not to be confused with* ASTROLOGY *—which lacks any kind of scientific basis and encompasses the belief that the* ***heavenly bodies*** *influence human lives.*

ancient stargazers kept a watch for omens

yes, this is definitely the center of things

INTRODUCING THE UNIVERSE

* The popular concept of what the universe is like has altered radically over the ages. At first people put the Earth at the center of the universe. Then they found that the Earth was a planet, and thought that the sun was the center. Then they realized the sun was an ordinary star, which was part of a great star island, or galaxy, and thought that this galaxy was the universe. Then they found that this galaxy was just one of many, and that many galaxies made up the universe. And that is about where we are today.

our sun and moon aren't the only ones in our corner of the universe

Just visible

The most distant galaxy we can see from Earth without the aid of a telescope or binoculars is the Andromeda Galaxy.

OUR UNIVERSE

* Continuing this trend of astronomical upsizing, the latest speculation among astronomers is that maybe the universe we know is only one of many—and that there may be many universes, making up a vast SUPERUNIVERSE.

* *Our corner of the universe is dominated by the sun. It is a star, like the other stars in the sky only very much closer. Traveling with the sun through space are nine planets: Mercury, Venus, Earth, Mars, Jupiter, Saturn, Uranus, Neptune, and Pluto, in order of distance from the sun.*

Most of these planets have smaller bodies—MOONS—circling around them. Numerous smaller bodies—ASTEROIDS and COMETS—also form part of the sun's family, or SOLAR SYSTEM.

THE SUN AND STARS

* The sun is a very ordinary kind of star, of average size and average brightness. Astronomers know of stars that are smaller and dimmer, and ones that are bigger and brighter. ***All stars are great globes of intensely hot gas, and pour out prodigious amounts of energy into space, in the form of heat, light, and other*** RADIATION.

* ***Together, the sun and the other stars we can see in the sky form a great star island, or*** GALAXY, ***in space. Astronomers estimate that our galaxy contains about 100 billion stars. It is so big that it would take a beam of light 100,000 years to travel from one side to the other.***

GALAXIES

* *Our galaxy forms part of a group of about 20 galaxies traveling through space together. On a grander scale, this group forms part of a bigger cluster of thousands of galaxies. And on the biggest scale (at least, as far as we know today), many such clusters make up the universe.*

Interstellar tour

Suppose we set off from Earth on a tour of the universe, in a spaceship that can travel at the speed of light:

* *We would reach the moon in less than 1½ seconds.*
* *Venus, the nearest planet, would flash past in about 2½ minutes.*
* *In 8½ minutes we would be zooming around the sun, en route for the most distant planet Pluto, which we would encounter in a little over 5½ hours.*
* *It would take 4¼ years, though, to reach Alpha Centauri, the nearest bright star.*
* *It would take us nearly 30,000 years to reach the center of our galaxy.*
* *It would take us about 2,300,000 years to reach the Andromeda Galaxy (the most distant object we can see with the naked eye).*
* *It would take us more than 13,000,000,000 years to reach the farthest objects visible in our most powerful telescopes.*

GET OBSERVING!

* Having introduced our universe of galaxies and stars, planets and moons, and other bits and pieces hurtling through space, let's now see how astronomers gather knowledge about what exists and what happens in the heavens. At the basic level, they just use their eyes. So get going! Start observing tonight!

KEY WORDS

DARK ADAPTATION: letting your eyes adjust to the dark (once they've done so, you'll see more stars)

PLANISPHERE: a kind of movable star map that shows the view of the heavens at a specific latitude for every hour of every night of the year

ADAPTING THE EYE

When you are ready to start stargazing, don't expect to see the heavens aflame from the word go. **Give your eyes time to adjust to the darkness.** This is called "**dark adaptation,**" and it is brought about by physical changes in the eye. For one thing, in the dark the pupil opens up to its fullest extent—up to almost ¼ inch. For another, a fluid called **visual purple** floods the retina, making it infinitely more sensitive than it is in daylight.

A GOOD SITE

* Ideally, it's best to go stargazing in the depths of the countryside, ***far away from the glare of street lights and car headlights and away from polluted city skies***. The author would particularly recommend a trip to the Australian outback, where he spent March 1986 observing HALLEY'S COMET.

There, in perfect darkness, the stars are startlingly brilliant, and the MILKY WAY seems ablaze.

★ ***But if you can't get to the countryside, let alone the outback, don't worry. Most urban areas have their darker spots where the*** CONSTELLATIONS ***and the brighter planets are easily visible.***

it's easier to find your way around the heavens if you've got a planisphere and the right kind of flashlight

GET SET

★ ***It's worth making a few preparations before you go observing.*** The first essential is to bundle up. Even in summer, nights can be cool. And in winter, when dark skies make for better viewing, thick sweaters, parkas, scarves, gloves, and balaclava helmets are the rule. And don't forget a couple pairs of socks, plus warm (ideally fur-lined or fleece-lined) boots, to insulate your feet from the frozen ground.

bundle up when out at night

★ You will need a STAR MAP to help you find your away around the constellations. A PLANISPHERE is useful too, for a quick check on your celestial bearings. To read these, use a flashlight with a red bulb or filter to give a subdued light, or a special STARGAZING FLASHLIGHT with a low light level.

THE PLANISPHERE

A planisphere is a device with movable disks, the bottom one carrying a star map. **When you match time and date scales on the disks, a window in the upper one shows you a plan of the heavens above you.** Planispheres are specific to different latitudes and to the Northern and Southern Hemispheres. **But for the given latitude, they show views of the heavens for every hour of every night of the year.**

SEEING FAR

★ We can see quite a lot in the heavens with the naked eye—the constellations, planets, meteors, comets, and the moon—in the same way that our ancestors did (though in our brighter modern world, not as distinctly). Using a telescope, however, we can see a lot more.

there's plenty to see with the naked eye

KEY WORDS

MOUNTING: a support for a telescope that allows it to swivel up and down and from side to side to follow the stars

REFLECTOR: a telescope that uses mirrors to gather and focus light

REFRACTOR: a telescope that uses lenses to gather and focus light

but eyes aren't as good as telescopes!

POOR EYES

★ As an astronomical instrument, the human eye is not really very good. Its great disadvantage is that it has a small opening, or aperture—the pupil. And this limits the eye's LIGHT-GATHERING POWER. It also limits the RESOLVING POWER of the eye—its ability to separate stars that are close together.

REFRACTORS AND REFLECTORS

★ *A telescope is a much better light-gatherer and also has better resolving power. It produces a magnified image, but reduces the field of view—the area of sky you can see.*

★ There are two main kinds of telescope, the REFRACTOR and the REFLECTOR. *The refractor is so called because it uses lenses, which refract (bend) light. The reflector uses mirrors, which, of course, reflect light.*

MOUNTING UP

★ ***Telescopes must be sturdily supported, so they don't wobble and blur the image.*** Most amateur telescopes are supported on tripods. Big observatory telescopes have massive supports. But whatever their size, telescopes need to be mounted in such a way that they can be swiveled to observe different parts of the sky.

★ ***There are two main methods of mounting a telescope.*** The simplest is the ALTAZIMUTH, which is like the "pan-and-tilt" head of a camera tripod. With this mounting, the telescope can be moved up and down (IN ALTITUDE) and swiveled horizontally (IN AZIMUTH). But there is a snag. ***If you want to follow a particular star, you have to move the telescope in altitude and azimuth continuously as the stars wheel across the sky.*** An equatorial mounting (see right) solves this problem.

telescope with an equatorial mounting

EQUATORIAL MOUNTING

Following the stars across the sky is easier with an **equatorial mounting**. You mount the telescope on two axes at right angles to each other. One of the axes **(the polar axis)** is lined up so it is parallel with the Earth's axis. When the telescope is moved around this axis, it follows the same path as the wheeling stars. The telescope can also swivel around the other axis **(the declination axis)** to locate stars high or low in the sky.

Good sizes

Amateur astronomers use both refractors and reflectors. Useful sizes are 2–3 inch for a refractor and 6–8 inch for a reflector. ***The figures refer to the diameter of the light-gathering lens or mirror.***

the first telescopes had glass lenses

THROUGH A GLASS, BRIGHTLY

★ The first telescopes were refractors, made with glass lenses. Galileo **pioneered** telescopic astronomy **using one in the early 1600s. Once their defects had been overcome, telescopes opened up a whole new universe for astronomers.**

THE X-FACTOR

Telescopes—reflectors as well as refractors—usually come with two or three eyepieces, which give different magnifications of the image—such as x 20 (magnifying 20 times), x 75, and x 200. You would select a high magnification (with a limited field of view) to study, say, part of the moon in detail. But you would find it better to use a low magnification (with a wider field of view) to appreciate the beauty of a star cluster.

THE OBJECT

★ *A refractor has two lenses.* At the front is the larger one, called the OBJECTIVE or OBJECT GLASS. At the rear is the lens you look through, called the EYEPIECE or OCULAR.

a refracting telescope

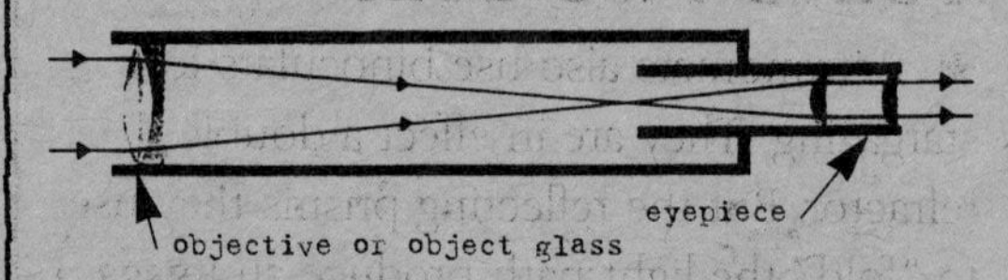

★ *The object glass gathers incoming light and brings it to a focus as a sharp image, which is magnified by, and viewed through, the eyepiece*. The body of the telescope is, well, telescopic; and the eyepiece slides in and out of the body tube for focusing.

telescopes make the moon and other things appear upside-down

UPSIDE-DOWN MOON

★ *Both refracting and reflecting telescopes produce an inverted (upside-down) image*. For most astronomical

with binoculars (unlike telescopes), birds and planes fly the right way up

purposes, this doesn't matter—stars look the same any way up! But it does matter when you are observing the moon or a planet, since in a telescope you are viewing them with south at the top and north at the bottom. ***To make life easier, many moon maps are often printed this way up.***

USING TWO EYES

★ Astronomers also use binoculars for stargazing. They are in effect a double refractor. But the reflecting prisms they use to "fold" the light path produce an image that is the right way up. This means you can use them for watching birds or planes, without the birds or planes appearing to fly upside-down. ***A useful size for astronomical viewing is 7 x 50. The 7 refers to the magnification, the 50 to the diameter of the object lens in millimeters. For serious sky study, 25 x 105 binoculars can be used, but they need to be mounted on a tripod or clamped to some other form of support because of their weight.***

KEY WORDS

EYEPIECE:
the viewing lens of a telescope

OBJECTIVE OR OBJECT GLASS:
the light-gathering lens of a refractor

The Yerkes refractor

The world's biggest refractor—at Yerkes Observatory, in Wisconsin—was completed in 1897. It has an objective that is nearly 40 inches across. *Big refractors are no longer built due to the difficulty of supporting their lenses. Because they can only be supported around the edges, the lenses tend to distort under their own weight. Another snag is that the thicker the lenses are, the more light they absorb.* ***Reflectors don't have these drawbacks, which is why all the world's biggest telescopes are reflectors, not refractors.***

they'll never see this

DOING IT WITH MIRRORS

* Reflecting telescopes that use mirrors suffer from fewer inherent defects than refracting telescopes, and have the advantage that they can be built in much larger sizes. The latest big ones have such prodigious light-gathering power that they could, it is said, spot a candle burning on the moon.

Big eyes

The 200-inch Hale reflector at Palomar Observatory, near Los Angeles, is the largest successful ***single-mirror telescope*** *in the world. But using lots of little mirrors to make one big mirror allows larger reflector telescopes to be built. Each of the twin Keck telescopes at Mauna Kea Observatory, in Hawaii, has a mirror 33 ft across. The mirror is built in 36 separate hexagonal segments, each easily supported and free from distortion. Computer-controlled actuators align the segments so that the whole mirror is always the perfect shape. This technique is known as* ***adaptive optics****.*

CURVED COLLECTORS

* ***In a reflecting telescope, light is gathered at the bottom of the telescope tube by a concave dish-shaped mirror.*** This mirror, called THE PRIMARY, is shaped like a paraboloid—a surface that has the cross-sectional shape of a parabola.

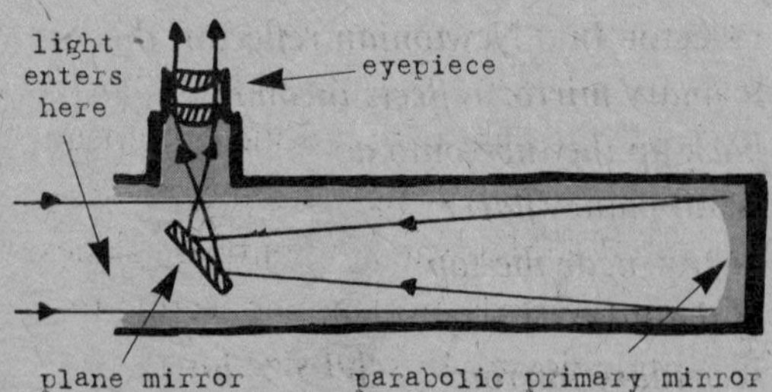

the Newtonian reflector

* ***The primary mirror can focus incoming light in a variety of ways and in different places.*** In many observatory telescopes, the primary simply focuses the image onto a photographic plate located

near the top of the telescope tube. This is known as the PRIME-FOCUS POSITION. Alternatively, a SECONDARY MIRROR in a similar position reflects the image back down the tube through a hole in the primary mirror onto a photographic plate or instruments beneath. This is called the CASSEGRAIN FOCUS. But ***the most common system used by amateur astronomers is*** THE NEWTONIAN.

Isaac Newton built a little reflector like this

THE NEWTONIAN

* The Newtonian reflector is named after SIR ISAAC NEWTON because it uses the same mirror system as in his original reflector. ***In a Newtonian reflector, the primary mirror reflects incoming light back up the tube onto a small plane (flat) mirror near the top of the tube. This secondary mirror reflects a focused image into an eyepiece set in the side of the tube at a convenient height for viewing.***

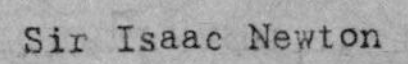
Sir Isaac Newton

ESO's VLT

The status of the twin **Keck telescopes** as the most powerful eyes on Earth will not last long. Four identical telescopes, each with a single mirror 27 ft in diameter, will shortly come into operation in Chile, at the **European Southern Observatory (ESO)**. Together they make up the **Very Large Telescope (VLT)**. The first of the four saw "first light" in May 1998. When all four come on stream, they should give the same superlative clarity of vision as the Hubble Space Telescope.

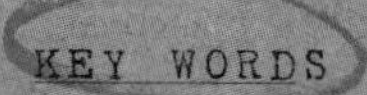
KEY WORDS

NEWTONIAN: REFLECTOR
a reflector that uses the same mirror arrangement as Newton's original reflector

CAPTURING THE IMAGE

* Professional astronomers seldom look through the big telescopes they have at their disposal. Instead, they use their telescopes as huge cameras, and take pictures of the sky that they can later peruse at leisure. Or, more and more, they use modern electronics to capture images.

photographs can provide a handy permanent record of your observations

Bond's breakthrough

US astronomer William Bond (1789–1859) was one of the pioneers of **astrophotography**. *In 1850 he produced the first photographs of stars—the same year in which he discovered the innermost of Saturn's three main rings.*

KEY WORDS

CCD:
microchip that is light-sensitive

EMULSION:
the layer on a photographic film that contains light-sensitive chemicals, usually silver salts

ON FILM

* *Film has a great advantage over the human eye in that it can in effect store light in its emulsion. The longer you expose it, the more light it stores*. So if an astronomer exposes film in his telescope for long periods, faint objects invisible to the eye become visible. A dusk-till-dawn exposure will store the maximum starlight, and can capture images of stars and galaxies that lie millions of light-years away.

* The light gathered by a telescope is not only directed onto photographic film but is also fed to other instruments, such as PHOTOMETERS and SPECTROMETERS. *Photometers measure the brightness of celestial objects. Spectrometers split up the incoming starlight into a spectrum for study. Star spectra hold the key to our understanding of the stars*.

MICROCHIP MAGIC

★ ***Astrophotography is on its way out, because astronomy, like other scientific fields, is making good use of microchips.*** The ones astronomers use—called CCDs or CHARGE-COUPLED DEVICES—are photosensitive (sensitive to light). Today's video cameras use the same kind of chip.

★ ***The beauty of using CCDs is that they are much more sensitive to light than photographic emulsion is, so faint objects in the heavens can be captured in a relatively short exposure.*** Exposure to light sets up tiny electric charges on the thousands of tiny picture elements, or pixels, on the CCD. These charges are

astronomers sometimes use computers to enhance or "massage" images

converted to electronic signals that can be fed into a computer and displayed as a picture on a screen. ***The signals can be "massaged" in all kinds of ways to produce realistic or false-color images and to highlight particular features.***

telescopes have long doubled as cameras

DON'T BLINK NOW!

Astronomers searching for new objects in the heavens use a device called a blink microscope to compare photographs. Two photographic plates of the same part of the heavens, taken at different times, are rapidly alternated under a viewing eyepiece. If anything has changed its position or has appeared, it will therefore show up. Variable stars and asteroids are among the objects that can be detected by this technique.

MOUNTINGS IN THE MOUNTAINS

★ The world's finest optical observatories are built high up in the mountains, usually in a dry climate. There the heavens can be viewed with greater clarity than is possible at lower altitudes, enjoying cloudless skies for most of the time.

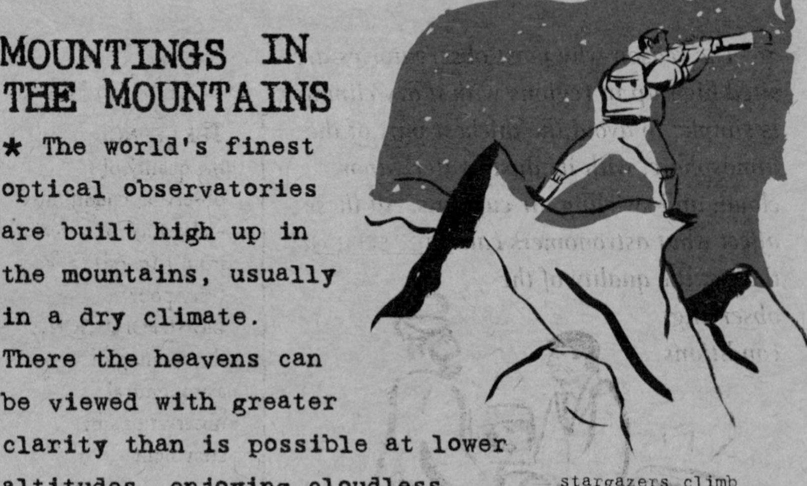

stargazers climb mountains for a clearer view of the skies

POLLUTING LIGHTS

The Hale telescope at Mount Palomar Observatory is being increasingly affected by light from the mushrooming suburbs of nearby Los Angeles. In fact **light pollution is becoming a real problem worldwide, especially for amateur astronomers**. Russian proposals to launch huge mirrors into space to illuminate their northern territories have alarmed astronomers—a big shiny mirror in the night sky will overwhelm the light of the stars.

ON TOP OF THE WORLD

★ The twin Keck reflectors in Hawaii are sited at Mauna Kea Observatory, along with other fine instruments, including the UKIRT (United Kingdom Infrared Telescope). The Observatory is built at an altitude of 13,800 ft, on the summit of an ancient volcano. This is appropriately named Mauna Kea ("White Mountain") because it is often snowcapped, despite being in the tropics.

★ Other observatories located high up include the famous Mount Palomar Observatory near Los Angeles, Kitt Peak National Observatory in the Quinlan Mountains of the Arizona Desert, the Roque de Los Muchachos Observatory on La Palma, in the Canary Islands, and La Silla Observatory in the Andes, in Chile.

★ *The reason why most observatories are sited high up in regions with a dry climate is simple: to avoid the thickest part of the atmosphere, with its dust, water vapor, cloud, and swirling air currents. All these affect what astronomers call* THE "SEEING," ***that is, the quality of the observing conditions.***

observatories are now often international

SHARING

★ Astronomy has always been cosmopolitan in its outlook, and it is becoming even more so, both at the professional and the amateur level. ***This not only means exchanging information when, say, a new object appears in the sky, but extends to sharing the costs of constructing and staffing new telescopes and observatories, which can be mighty expensive.*** For example, several countries have financed the building of the Very Large Telescope at the European Southern Observatory, in Chile, which is due to come into full operation in the year 2000.

KEY WORDS

THE SEEING:
the quality of observing conditions —classified on a scale from I (good) to V (very poor)
LIGHT POLLUTION:
the spoiling of astronomical observations by stray light

Radio observatories

Unlike ordinary optical observatories, radio observatories (see p.22) can, within reason, be sited anywhere. **Weather conditions seldom affect the radio signals streaming in from outer space.** *This allows Britain's famed Jodrell Bank Observatory to operate perfectly well from a ground-level site near Manchester—where, according to vaudeville comedians, it always rains.*

...and it's raining again in Manchester

TUNING IN TO THE HEAVENS

* The study of the radio waves reaching us from the heavens has revolutionized our knowledge of the universe, leading to the discovery of remarkable bodies such as quasars and pulsars. Radio "pictures" give us a view of the heavenly bodies quite unlike that of ordinary photographs.

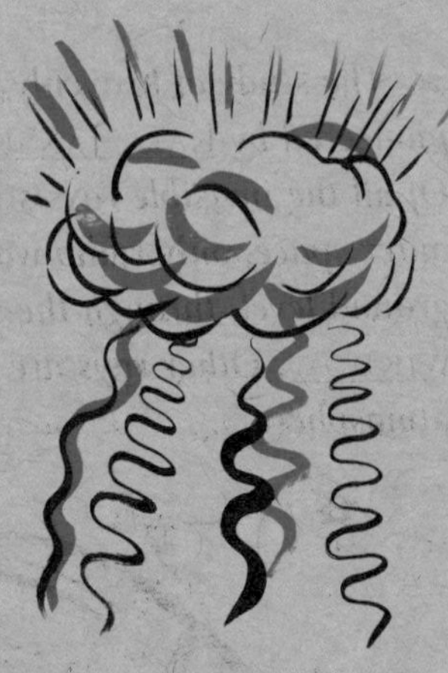

tuning in to the heavens can be a revelation

KEY WORDS

ANTENNA: another word for aerial (a device for picking up radio waves)

FALSE COLOR: color that is not true to life

RADIO ASTRONOMY: the study of the radio waves given out by heavenly bodies

THE RADIO WINDOW

* During their lengthy night-time vigils, astronomers study the faint light that reaches us from the heavens. ***But by studying light alone, astronomers can't get a true picture of the universe. This is because light is not the only kind of radiation that stars and galaxies give out.*** They give out many other kinds, too, which we cannot see.

KARL JANSKY (1905-50)

In 1931, while investigating sources of radio interference for Bell Telephone Laboratories at Holmdel, New Jersey, American radio engineer Karl Jansky discovered that the heavens give out radio waves. The aerial he used to pick up signals consisted of wire hoops supported on a wooden frame that could rotate on a set of wheels taken from a Model T Ford.

Jansky building his telescope on wheels

★ *The study of heavenly radio waves was pioneered by* KARL JANSKY *in the 1930s. Of all the invisible rays streaming in from outer space, only radio waves reach us at ground level, through the so-called* RADIO WINDOW. *Other rays are blocked by the atmosphere.*

big dishes aren't just for TV

BIG DISHES

★ *Radio astronomers pick up radio waves from outer space with specially designed telescopes. These usually take the form of huge metal dishes with a central aerial that gather and focus the incoming radio signals.* After being amplified billions of times, the signals are fed into a computer, or recorded on magnetic tape. The computer is able to analyze the signals and display them as FALSE-COLOR RADIO "PICTURES"—equivalent to what we might see if our eyes were sensitive to radio waves.

GREAT COMBINATIONS

Even if it's as large as the one in Arecibo in Puerto Rico (which has a dish 1000 ft across) **a single radio telescope can't show much detail of a heavenly radio source. But a combination of several telescopes can.** At Socorro, in New Mexico, there is a powerful radio telescope known as the **Very Large Array (VLA)**. It consists of 27 dishes, each 82 ft in diameter, arranged in a Y-shaped pattern. Each arm of the "Y" is 13 miles long, and the dishes can be moved along each arm. **When signals received from all the dishes are combined, they produce a radio picture equivalent to one from a dish 17 miles across and show as much detail as in a light picture.** On a much larger scale, it's possible to combine radio telescopes across the globe. **Under optimum conditions, this technique—known as VLBI (Very Long Baseline Interferometry)—could in effect create a dish the size of the Earth.**

TELESCOPES IN ORBIT

* In recent years, the importance of radio astronomy has been rivaled by satellite astronomy. By lofting instruments high into space, astronomers can study invisible rays from outer space (such as gamma rays and microwaves) that are blocked by the Earth's atmosphere.

satellites have opened up whole new areas of astronomical observation

KEY WORDS

ABSOLUTE ZERO: the lowest possible temperature (–459.69°F)

COSMOLOGIST: astronomer who studies the origin and evolution of the universe

ELECTROMAGNETIC WAVES: invisible electromagnetic disturbances in space

SATELLITE/ARTIFICIAL SATELLITE: manmade moon that orbits around the Earth

SPEEDY MOVERS

* Powerful rockets are needed to boost satellites to a speed at which they can beat Earth's gravity and climb into space. *The minimum speed required (orbital velocity) is a staggering 17,400 mph, or more than 30 times the speed of a jumbo jet. At such a speed, a satellite can go into orbit around the Earth at a height of a few hundred miles.*

* What kind of instruments a satellite carries depends on what it is designed to do. *Detecting different forms of radiation requires different types of equipment. The instruments of* INFRARED TELESCOPES, *for example, are cooled by liquid helium to within a few degrees of absolute zero in order to make them ultrasensitive to faint cosmic sources of* INFRARED RADIATION, *or heat.*

THE FIRST EXPLORER

★ *The Russians pioneered the Space Age when they launched* Sputnik 1 *on October 4, 1957. But it was the first US satellite—*Explorer 1, *launched on January 31, 1958—that pioneered satellite astronomy by making the first discovery of the Space Age*. With instruments devised by the US scientist JAMES VAN ALLEN, *Explorer 1* discovered two huge doughnut-shaped regions of intense radiation around the Earth, which became known as THE VAN ALLEN BELTS.

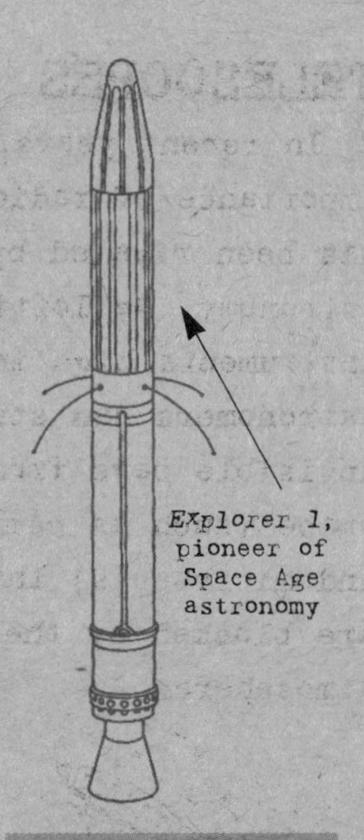

Explorer 1, pioneer of Space Age astronomy

"BIG BANG" RIPPLES

★ Of particular interest to cosmologists was the discovery made by the US satellite COBE (Cosmic Background Explorer), launched in 1989. Surveying the whole sky at microwave wavelengths, COBE found slight variations (ripples) in the background radiation left over from THE BIG BANG, the gigantic explosion that created the universe. *Cosmologists had theorized for some time that there would be ripples like these.*

THE ELECTRO-MAGNETIC SPECTRUM

The invisible rays given off by stars and galaxies belong to the same family of rays as light rays. They are all **electromagnetic waves**, and form what is called the **electromagnetic spectrum**. The crucial difference between them is that they have different wavelengths. **In order of increasing wavelength, the waves include gamma rays, X rays, ultraviolet rays, visible light, infrared rays, microwaves, and radio waves.**

I'm just astounded

THE HUBBLE REVOLUTION

★ Up in orbit, high above the atmosphere, is the most powerful optical telescope the world has ever known—the Hubble Space Telescope (or HST). Able to view the universe with perfect clarity, it is sending back images that are astounding astronomers.

the HST—"a new window on the universe"

HUBBLE DATA

Launch date: April 25, 1990
Orbit: 380 x 385 miles
Size of primary mirror: 7.9 ft
Length: 46.6 ft
Overall width: 39.3 ft (across solar panels)
Diameter: 14 ft
Weight: 11 tons

BLURRED VISION

★ ***Built at a cost approaching 1.5 billion dollars, the Hubble Space Telescope was launched into orbit by the space shuttle*** **Discovery** ***in April 1990***. It should have been launched four years earlier, but the launch was postponed when the shuttle fleet was grounded after the *Challenger* disaster in January 1986. After exhaustive systems checks, Hubble scientists waited eagerly for "first light" (the time when the telescope would begin sending back images).

★ ***First light occurred on May 20, 1990—but with it came a bombshell. The images were blurred! And no matter how skillfully the scientists manipulated the incoming electronic image data, they remained blurred***. Subsequent investigation revealed that, because of a manufacturing error, the curvature of the main light-gathering mirror was inaccurate.

The curvature was out by only about one-fiftieth of the width of a human hair—but this was enough to blur the images, making them little better than those that can be obtained from large Earth-based telescopes. Computer-massaging of the data improved the images, but not much.

TRY WEARING GLASSES

★ *After much agonizing about how to fix the problem, the National Aeronautics and Space Administration (NASA) decided on what might be called the optician's solution. If you find you have blurred vision, wear glasses*. At least, that is what they planned for one of the main instruments, the wide-field PLANETARY CAMERA designed to take detailed images of the planets. *To correct the vision of the other main instruments, the* FAINT-OBJECT CAMERA *(for imaging faint stars and galaxies) and two* SPECTROGRAPHS *(for recording stellar spectra)*, NASA assembled an arrangement of mirrors in a unit called COSTAR (short for Corrective Optics Space Telescope Axial Replacement).

★ In December 1993 shuttle astronauts began a series of lengthy spacewalks in which they installed the new equipment and replaced solar panels and sets of gyroscopes that had malfunctioned. *Within weeks the HST was sending back superlative images and fulfilling its prelaunch promise of opening up a new window on the universe*.

Hubble feats

* *Records scars on Jupiter created by the impact of comet Shoemaker Levy 9.*
* *Finds* **"pillars of creation"** *in the Eagle Nebula where stars are forming.*
* *Pinpoints* **protostars** *with* **proplyds**—*enveloping disks in which planets might be forming.*
* *Images gas clouds blasted into space from supernovae.*
* *Peers into the center of active galaxies.*
* *Discovers a "deep field" of distant galaxies never seen before.*

an extraordinarily tricky repair job

PROBING THE SOLAR SYSTEM

* Astronomers not only observe the heavenly bodies using Earth-orbiting satellites, they also send space probes to visit them *in situ*. The first probes were despatched to the moon in 1959. Since then the moon and all the planets, except Pluto, as well as comets and asteroids, have been subjected to close scrutiny from these robot explorers.

Highlight discoveries

** Mercury has a cratered surface like the moon. * Venus is hotter than an oven. * Rivers once flowed on Mars. * Jupiter's moon Io has volcanoes. * Winds on Saturn can blow at up to 1,000 mph * Uranus's moon Miranda was once smashed to pieces. * Neptune's moon Triton has geysers. * Halley's comet is about 10 miles across and is shaped like a potato.*

KEY WORDS

ESCAPE VELOCITY: The speed a space probe has to reach to escape from the grip of Earth's gravity

ON TARGET

* At a speed of 17,400 mph a satellite can climb into space but, once there, it is still held in the firm grip of Earth's gravity. To escape completely from gravity, a space probe must travel even faster, at a speed of at least 25,000 mph. This speed is called the ESCAPE VELOCITY. ***But achieving the right speed is just the beginning of space scientists' problems***. They have to make sure the probe is on exactly the right TRAJECTORY (path), so that it will arrive at a point in space at the same time as its target, which is moving in orbit around the sun. ***Thanks to the sophisticated computers now available, trajectories can be worked out with extreme accuracy, so that probes can be flown close to the target or made to land on it, after a journey of maybe billions of miles.***

HIGHLIGHT MISSIONS

Probe	Launched	Target/mission
Luna 2	September 1959	Pass by the moon
Mariner 4	November 1964	Mars
Surveyor 1	May 1966	Land on the moon
Venera 4	June 1967	Venus
Lunokhod 1	November 1970	Land a vehicle on the moon
Mariner 9	May 1971	Orbit Mars
Pioneer 10	March 1972	Jupiter
Pioneer 11	April 1973	Jupiter, Saturn
Mariner 10	November 1973	Mercury
Viking 1	August 1975	Orbit and land on Mars
Viking 2	September 1975	Orbit and land on Mars
Voyager 2	August 1977	Jupiter, Saturn, Uranus, Neptune
Voyager 1	September 1977	Jupiter, Saturn
Pioneer-Venus 1	May 1978	Orbit Venus
Giotto	July 1985	Observe Halley's Comet
Magellan	May 1989	Orbit Venus
Galileo	October 1989	Jupiter
Ulysses	October 1990	Sun
SOHO	December 1995	Sun
Global Surveyor	November 1996	Orbit Mars
Pathfinder/Sojourner	December 1996	Land a vehicle on Mars
Cassini-Huygens	October 1997	Saturn, Titan

DOING THE JOB

* Probes carry telescopes, detectors to pick up radiation, cameras, and all sorts of other instruments, to collect information about their targets. Those bound for the nearer planets use solar panels to generate electricity for their instruments, radio, and spacecraft systems. The ones to Jupiter and beyond are nuclear-powered by RTGs (radioisotope thermoelectric generators).

Snowflake power

The energy in the radio signals sent back from probes traveling to the outer planets is a fraction of that in a falling snowflake.

both *Voyagers* surfed Saturn's rings

A GRAND TOUR

* In the late 1970s the five outer planets from Jupiter to Pluto were coming up to an alignment that would not be repeated for another 176 years. NASA had worked out a plan by which spacecraft could be launched to visit all of them, and named it the Grand Tour.

THE TWO VOYAGERS

This ambitious concept was only possible using a technique called GRAVITY-ASSIST, which used the gravity of one target planet to accelerate the probe toward the next. * In the event, a more modest project, named VOYAGER, was approved. ***Two identical craft would be launched. The first would explore Jupiter and Saturn, but the second would travel on to explore Uranus and Neptune as well.***

DEEP-SPACE VOYAGER

One space probe more than any other has revolutionized our knowledge of our solar system—and that is *Voyager 2*. On a 12-year journey of discovery, it visited all four of the giant planets before entering interstellar space and heading off towards far distant stars.

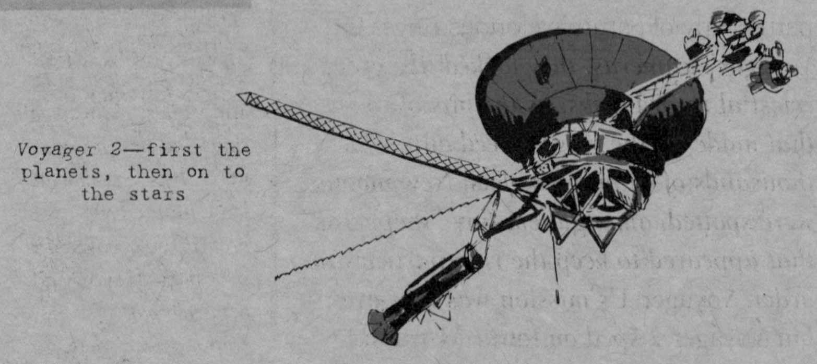
Voyager 2—first the planets, then on to the stars

BY JOVE

* *Voyager 2* was in fact launched first, in August 1977, with its sister craft following a month later. But *Voyager 1* arrived at Jupiter first, in early March 1979, with *Voyager 2* arriving in late April. ***Nothing discovered by the earlier Pioneer probes prepared mission scientists at the Jet Propulsion Laboratory in California for the feast of data and staggering images sent back by the Voyagers. Jupiter's surface presented a kaleidoscope of vivid colors, showing the turbulent currents rippling through the cloudy atmosphere. The Great Red Spot looked awesome. The Jovian moon Io was seen spewing out gases from erupting volcanoes. Other discoveries included a ring around the planet and several tiny new moons.***

VIA SATURN

* The slingshot effect of Jupiter's gravity accelerated the two Voyagers toward their next planetary encounter, with Saturn. In their close-up investigations of the planet, they again returned stunning images, particularly of Saturn's glorious rings. ***In Voyager's cameras, they looked like great celestial record disks, as the bits of rock that made up the rings traced out thousands of shining ringlets. New moons were spotted, among them tiny "shepherds" that appeared to keep the ring particles in order.*** **Voyager 1*'s mission was now over, but* Voyager 2 *sped on toward Uranus.***

Battered moon

In January 1986 Voyager 2 *took the first ever close-up pictures of Uranus and its moons and faint rings. Here, too, new moons were discovered but the biggest surprise was the battered surface of Miranda, which was like nothing ever seen before.*

Triton proved to have a whole lot of geysers

Last stop for *Voyager 2*

Voyager 2*'s last port of call (in August 1989) was Uranus' twin planet, Neptune, which surprised everyone with its cloudy weather and deep-blue atmosphere. Once again rings and new moons were observed, and the large moon Triton proved to be covered with pinkish snow and peppered with geysers erupting everywhere.*

CHAPTER 1

HOW IT ALL BEGAN

★ We need to travel back in time some five millennia to trace the beginnings of the science of astronomy. It grew up in the early civilizations of the Middle and Far East. The early priest-astronomers chronicled comings and goings in the heavens, developed accurate calendars, and were able to predict eclipses. But they had no idea what the universe was like.

KEY DATES

C. 4000 BC
Chinese astronomers record eclipses.

C. 2500 BC
Great Pyramid constructed.

C. 2100 BC
Stone circles of Stonehenge erected.

Ancient cosmology

Ancient civilizations had different ideas about what the universe was like. Early Indian priests thought it rested on huge pillars or on the back of an elephant. The Egyptians saw the Earth as being spanned by the star-spangled body of the sky goddess Nut, whose boats transported their sun god Ra across the heavens by day.

STARGAZING

★ Obviously, people started stargazing long before the early civilizations grew up. But, since they couldn't write, they couldn't leave any records. In any case, early prehistoric peoples were undoubtedly too continually concerned about where the next meal was coming from to pay much attention to celestial happenings. ***Only when farming developed, in about 8–10,000 BC, did they need to tap into the rhythm of the seasons in order, for example, to make sure they planted their crops on time.***

ancient astronomer

BETWEEN THE RIVERS

* The early civilizations we know as Babylonia and Assyria grew up in the fertile region called Mesopotamia (meaning "between rivers"), between the Tigris and Euphrates Rivers. ***They left astronomical records, from about 3000 BC, some depicting the same signs of the zodiac astrologers use today***. Like their southerly neighbors, the Chaldeans, the Babylonians studied the heavens as much to learn the will of the gods as for practical purposes. Thus were laid the foundations for the pseudoscience of astrology, which still has its adherents even today.

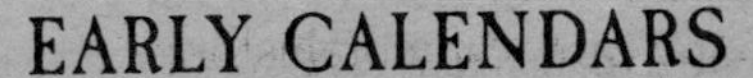

EARLY CALENDARS

* ***The Babylonians established a calendar based on the regular 29½ day cycle of the phases of the moon***. Lunar calendars don't fit in well with the changing seasons, but ***their Egyptian contemporaries in the Nile Valley developed an accurate solar calendar of 365 days, little different from the one we use today***.

* There is no doubt that the ancient Egyptians were skilled observers. ***The layout of their pyramids, for example, has been associated with the pattern of bright stars in the constellation Orion. And the Great Pyramid of Cheops is aligned with a star called Thuban, which at that time was the Pole Star and would have appeared virtually motionless in the sky.***

MEGALITHIC OBSERVERS

Far away in northwestern Europe, the peoples of Brittany and Britain were no dunces when it came to astronomical matters. This is witnessed by the rows of standing stones at Carnac, in Brittany, and above all by Stonehenge, on Salisbury Plain in southwestern England. This magnificent edifice has great stones set out in circular alignments that mark the summer solstice and other key times in the astronomical calendar.

THE GREEK ASTRONOMERS

* In ancient Greece, astronomy advanced by leaps and bounds. Philosophers pondered on the nature of the universe. Astronomers predicted eclipses, measured the size of the Earth, and put together star catalogs. One even had the temerity to suggest that the Earth traveled around the sun.

STARWATCHER

Perhaps the greatest Greek astronomer, **Hipparchus**, who lived around 150 BC, compiled a catalog of the positions of more than 800 stars. He also established the magnitude scale, still used to measure star brightness.

"Look, the Earth throws a curved shadow on the moon"

KEY DATES

585 BC Thales predicts an eclipse.
400s BC Meton works out his cycle for phases of the moon.
200s BC Aristarchus proposes a sun-centerd universe. Eratosthenes measures the circumference of the Earth.
C. 150 BC Hipparchus compiles his star catalog.
C. AD 150 Ptolemy writes the *Almagest*.

THE PERFECT SPHERE

* *The early Greeks inherited the flat-Earth concept of earlier civilizations, but later converted to the idea of a spherical Earth. This idea was well established by the time of* ARISTOTLE *(384–322 BC), who pointed out the proof: that the Earth threw a curved shadow on the moon during a lunar eclipse. In any case, the sphere was a perfect shape, and therefore appropriate for the Earth,*

it's all Greek to me

the early Greeks were excellent astronomers

which was, of course, the center of the universe. All the other heavenly bodies, from the sun to the stars, revolved in circles around the Earth.

* The Aristotelian idea of an Earth-centered universe was challenged some time later by an astronomer named ARISTARCHUS OF SAMOS (c. 310–230 BC). ***He suggested that the observed motions of the heavenly bodies could just as easily be explained if the Earth circled the sun, and not vice versa. But no one liked the idea much.***

"That's not exactly what I make it!"

MEASURING THE EARTH

* ERATOSTHENES OF CYRENE (c. 276–196 BC) knew that at noon on the summer solstice the sun shone down a well in Cyrene, some 500 miles from Alexandria. ***By measuring the angle at which the sun struck the ground in Alexandria at noon on that date, he was able—by geometry—to work out the Earth's circumference very accurately. He estimated it as 25,000 miles.***

Ptolemy

THE PTOLEMAIC UNIVERSE

In Alexandria, around AD 150, there lived an Egyptian astronomer named **Ptolemy. He was a skillful observer in his own right, but his principal claim to posterity rests in his seminal encyclopedic work summarizing the scientific knowledge of the ancient world.** It has come down to us via its Arabic translation and is known as the ***Almagest***—a name derived from the Arabic Al-Magisti, meaning "The Greatest." **As well as providing the only record of Hipparchus's star catalog, it sums up the classical Earth-centered view of the heavens, known as the Ptolemaic Universe.**

OUT OF THE DARKNESS

★ Following the collapse of the classical world, the period known as the Dark Ages began in Europe, when much of the knowledge of the ancient world was lost, or at any rate forgotten. Fortunately, Ptolemy's astronomical work was preserved in Arabic and inspired generations of Arabic astronomers. But it was not until centuries later that European astronomy began to assert itself once again.

KEY DATES

C. 820 The *Almagest* is translated into Arabic.
1054 Chinese astronomers witness a supernova in the constellation Taurus (the origin of the Crab Nebula).
1428 Ulugh Beigh founds observatory in Samarkand.
1543 Copernicus publishes book advancing the idea of a solar system.
1576 Tycho Brahe founds Uraniborg Observatory.

I wish I had an observatory

THE ARAB INFLUENCE

★ The *Almagest* was one of the learned works saved for posterity by scholars in a center of learning in Baghdad established in the early 800s by the famed Arab ruler Caliph Harun Al-Rashid. His opulent court and lifestyle were the inspiration behind the celebrated *Tales of the Arabian Nights*. An astronomical center grew up in Baghdad, followed by others throughout the Arab world. ***Arab astronomy flourished for six centuries, coming to an end with the death of*** ULUGH BEIGH ***(1394–1449), at the hand of his own son. Beigh set up an excellent observatory in Samarkand***, probably the finest there had ever been.

Harun Al-Rashid was a stimulating patron of the arts and sciences

astronomers who challenged the views of the church were persecuted

COPERNICUS' UNIVERSE

* In Renaissance Europe, philosophers and "natural scientists" began to question age-old beliefs. ***In astronomy, it was a modest Polish priest named*** NICOLAUS COPERNICUS ***(1473–1543) who rocked the boat. After studying his own and other astronomers' observations, he saw they could best be explained if the sun, not the Earth, was the center of the universe.***

* Copernicus knew that such thinking would be regarded as heresy by the Church, and might attract torture and even death. So he delayed publishing his views until he was on his deathbed. His book, *De revolutionibus orbium coelestium (Concerning the Revolutions of Celestial Spheres)*, marked the birth of modern astronomy.

Nicolaus Copernicus

Poor Bruno

Advocates of a ***Copernican universe*** *predictably suffered at the hands of the Church, none more so than* ***Giordano Bruno*** *(1548-1600). He was captured and tortured by the Inquisition, and finally burned at the stake in Rome in February 1600.*

TYCHO'S OBSERVATIONS

In 1572 a brilliant "new star" appeared in the constellation Cassiopeia and was studied closely by a Danish lawyer named **Tycho Brahe** (1546-1601). We now know it was a supernova—the death throes of a massive star (see page 95). It inspired Tycho to become a full-time astronomer, and **he later established a famous observatory, Uraniborg, on the island of Hven, in Denmark, where he made systematic observations of the heavenly bodies with unprecedented accuracy.**

KEY DATES

1609 Kepler publishes his first laws of planetary motion. Galileo pioneers telescopic observation.
1666 Newton formulates his law of gravitation.
1675 Greenwich Observatory founded in London.
1687 Newton's *Principia* published.

THE NEW AGE

* Brahe's meticulous observations proved a springboard to laws that explained how the planets moved and to the proof of a Copernican (sun-centered) universe. Observations using telescopes provided additional confirmation, while discovery of the laws of gravity explained why the heavenly bodies moved as they did.

Kepler

KEPLER'S LAWS

* Two years before Tycho Brahe died, a German mathematician named JOHANNES KEPLER (1571–1630) joined him as his assistant. He was able to see what Brahe, a confirmed Earth-centered man, could not; ***that the planetary positions so patiently observed by Brahe could only be explained***

THE *PRINCIPIA*

In 1687 Newton published his law of gravitation, along with many other innovative ideas, in what is often considered the most important scientific book of all time: *Philosophiae Naturalis Principia Mathematica*, usually known simply as the *Principia*. The book also gave an account of his pioneering work on optics and his laws of motion, and introduced a totally new form of mathematics, called calculus.

if the planets and the Earth traveled around the sun—and if they didn't travel in circles but in ellipses. This formed the basis of his three laws of planetary motion, the first two of which were published in 1609.

Newton discovers gravity

TWO GREAT GENIUSES

* ***That same winter an Italian named*** GALILEO GALILEI ***(1564–1642) put together a combination of lenses to make the first astronomical telescope. His eyes feasted on vistas no human eye had ever seen***. Galileo spotted mountains on the moon, satellites revolving around Jupiter, and the phases of Venus, and he marveled at the millions of stars packed together in the Milky Way.

* ***His observations confirmed that Copernicus had been right. In 1632 he published a book—his famous*** DIALOGUE—***ridiculing the Ptolemaic universe. Incensed, the Church leaders forced Galileo to recant his views and put him under house arrest for the rest of his life.***

Galileo

* The year Galileo died, another genius—ISAAC NEWTON (1642–1727)—was born. He studied "natural philosophy" (science and mathematics) at Cambridge University, but left the university in 1665, when the Great Plague struck England, and returned to his home at Woolsthorpe, in Lincolnshire.

That falling apple

Newton is supposed to have been prompted to thoughts about the nature of gravitation by a falling apple. *He reasoned that the Earth must be attracting the apple, and must also be attracting the moon and preventing it from shooting off into space. Over the next 20 years or so, Newton developed his ideas about such a force, which he realized must be an inherent property of all matter.* ***This led to his universal law of gravitation: that every body in the universe attracts every other body. The more massive and the closer they are, the greater is the attraction.***

KEY DATES

1672 Newton shows off his reflecting telescope.
1705 Edmond Halley predicts the return of the comet now named after him (see page 177).
1781 Herschel discovers Uranus.
1784 Charles Messier publishes his catalog of nebulae and other objects (see page 89).
1802 Herschel discovers binary stars.
1838 Friedrich Bessel measures the distance to a star (see page 68).
1846 Johann Galle discovers Neptune.

DISCOVERING NEW WORLDS

* As increasingly more powerful telescopes came into use, astronomers could probe deeper into the heavens. They found new worlds in our solar system, investigated double stars and clusters, and began to fathom what our galaxy was like.

NEWTON'S REFLECTOR

* In the course of his work on optics, Isaac Newton realized the drawbacks of the refracting telescopes that were coming into use. ***They were unsatisfactory because the lenses acted like prisms and color-blurred the image. So he decided to get around the problem by building a telescope with mirrors***. He presented his first reflector to the Royal Society in London in 1672 and was duly elected to that august body.

"And this is what I call a refractor..."

THE MUSICIAN'S PLANET

* ***It was more than a century later that a musician-turned-astronomer used a reflector to make a momentous discovery, which at a stroke doubled the size of the solar system***. He was German-born WILLIAM HERSCHEL (1738–1822), at the time living in Bath, in England. ***On March 13, 1781, he spied what he***

described as "a curious nebulous star or perhaps a comet," and watched it move slowly against the star field on the following nights. When the object's orbit was calculated, it proved to be neither a star nor a comet but a new—seventh—planet, which came to be called Uranus.

not a star, nor a comet—a planet!

PLANET HUNTERS

★ ***Astronomers eventually realized that Uranus wasn't following the orbit they expected, and yet another planet must be causing Uranus to move in such an irregular way.*** In 1845 an English astronomer, JOHN ADAMS (1819–92), calculated exactly where this new planet should appear in the sky and urged a search for it. But nothing happened.

★ The following year, French astronomer URBAIN LEVERRIER (1811–77) published his own predictions of where the new planet should be, which agreed with those made by Adams. ***Then on September 23, 1846 German astronomer Johann Galle (1812–1910) at the Berlin Observatory spotted the planet almost exactly where it had been predicted. This eighth planet was named Neptune.***

★ Calculations subsequently showed that even Neptune couldn't fully account for Uranus' irregular orbit. ***And so the planet hunt continued.***

THE STELLAR UNIVERSE

Astronomers had not been neglecting the stellar universe while all this planet hunting had been going on. William Herschel, though best known for discovering Uranus, spent most of his time studying the stars. It was he who first worked out that the Milky Way must be a view through the plane of our galaxy (see page 101), which he thought must have the shape of a convex lens. Herschel also was first to discover binary (double-star) systems.

THE EXPANDING UNIVERSE

★ By the mid-1800s most astronomers agreed that our galaxy was like a great bulging lens, and assumed that this was the universe. It was not until the 1920s that the truth was revealed: there are many galaxies beyond our own, and they are all racing away from us. The universe is expanding.

KEY WORDS

NEBULA:
a cloud of gas and dust between the stars

GALAXY:
an island of stars

KEY DATES

1905 Einstein publishes his special theory of relativity (see page 44).

1912 Slipher discovers receding nebulae.

1916 Einstein publishes his general theory of relativity (see page 45).

1917 The Hooker telescope comes into operation.

1925 Hubble discovers galaxies outside our own.

1930 Tombaugh discovers Pluto.

ROSSE'S SPIRALS

★ *Among mid-19th-century astronomical highlights was the discovery of the spiral nature of some of the nebulae.* In 1845 THE THIRD EARL OF ROSSE (1800–67) discovered the first spiral, M51, using a 72-inch reflector that he had built at Birr Castle, in Ireland. *Called the "Leviathan," it had a 59-ft-long tube and was the world's largest reflector at that time.*

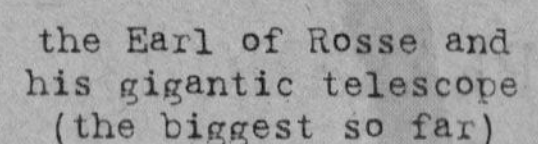
the Earl of Rosse and his gigantic telescope (the biggest so far)

SPEEDY SPIRALS

* In 1912 Vesto Slipher (1875–1969), working at the Lowell Observatory in Flagstaff, Arizona, became aware that ***nearly all the spiral nebulae he observed were hurtling away from us at high speed***.

NEW RESOLUTION

* In the 1920s another US astronomer, Edwin Hubble (1889–1953), began investigating nebulae using the world's biggest telescope, the 100-inch Hooker telescope at Mount Wilson Observatory. ***In 1925 he announced that he had been able to resolve (separate) individual stars in some nebulae—which should no longer be classed as nebulae, since they were galaxies in their own right***. By studying certain variable stars in them, he could estimate their distance. They proved to be hundreds of thousands of light-years away—far beyond the confines of our own galaxy.

PLUTO'S APPEARANCE

* Meanwhile, Slipher was continuing Lowell's search for a ninth planet. Lowell had worked out where a new planet ought to be, but never found it. In 1928 Slipher engaged a young astronomer, Clyde Tombaugh, to scan photographic plates for the planet. ***Examining plates taken in January 1930, Tombaugh found what he was looking for. The new planet came to be called Pluto***.

HUBBLE'S LAW

Hubble, too, found that virtually all the galaxies were rushing headlong away from us. In 1929 he announced that the speed of their retreat was directly related to their distance. The farther away they were, the faster they moved. This relationship, now known as Hubble's Law, provided convincing proof of an expanding universe and suggested a "Big Bang" origin for it.

Universal architecture

The basic architecture of the solar system and the universe was now in place. Since that time, modern telescopes have looked deeper into the universe and discovered many new kinds of heavenly objects. In fact, new discoveries are flooding in almost daily from the Hubble Space Telescope. As did the astronomer after whom it was named, it is opening up new vistas of our boundless universe.

THE EINSTEIN REVOLUTION

to Einstein, gravity was a "curvature" of spacetime

* At the beginning of the 20th century, a German clerk named Albert Einstein (1879-1955) introduced radical new ideas about space, time, motion, and gravity in two theories of relativity—the special theory (1905) and the general theory (1916). In mathematically challenging arguments, he demonstrated the equivalence of mass and energy and suggested that space is curved.

RELATIVITY

Einstein's special theory of relativity explored motion, time, the speed of light, matter, and energy. Here are some of its premises.

* All motion is relative, depending on where the observer is in relation to the moving object.
* Time is relative and passes faster or slower depending on the speed at which you are traveling.
* Nothing can travel faster than the speed of light.
* Mass (m) and energy (E) are related by the equation $E = mc^2$ (c being the speed of light).

TIME DILATION

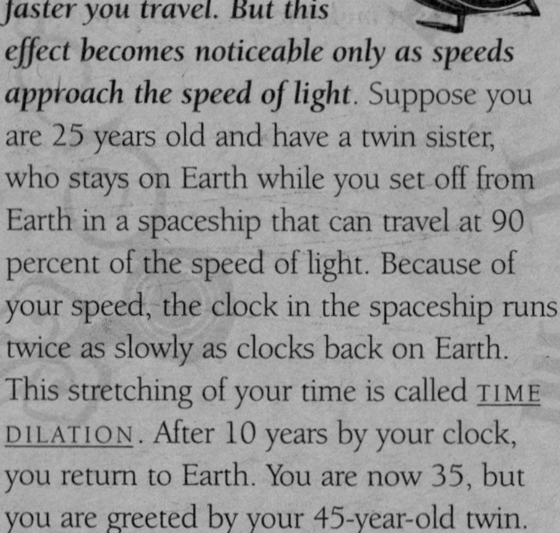

* ***The* SPECIAL THEORY OF RELATIVITY *states that time passes more slowly the faster you travel. But this effect becomes noticeable only as speeds approach the speed of light*.** Suppose you are 25 years old and have a twin sister, who stays on Earth while you set off from Earth in a spaceship that can travel at 90 percent of the speed of light. Because of your speed, the clock in the spaceship runs twice as slowly as clocks back on Earth. This stretching of your time is called TIME DILATION. After 10 years by your clock, you return to Earth. You are now 35, but you are greeted by your 45-year-old twin.

IN GENERAL

★ In his GENERAL THEORY OF RELATIVITY, Einstein introduced the idea of spacetime. Normally, we talk about three dimensions: height, length, and width. ***But because everything in space is constantly moving, another dimension (time) is needed to define an event or locate an object.***

★ ***The general theory looks at gravity in a different way. It is not a force as conventionally conceived. It is actually a distortion—a curvature—of spacetime caused by massive objects. Consequently, light and other radiation would be deflected from a straight path by the curving space in the vicinity of a massive object.*** And this is exactly what astronomers find, most spectacularly in the effect called GRAVITATIONAL LENSING. In this process, light from a distant quasar is bent by the gravity of an intervening galaxy, ***which in effect acts like a lens and produces a multiple image in photographs.***

THAT EQUATION

Einstein's equation $E = mc^2$ explains the huge energy output in nuclear-fusion reactions that take place in stars. When fusion between nuclei takes place, a certain amount of mass is lost. This lost mass represents matter that has been converted into energy—and because c^2 (the speed of light squared) is a huge quantity, the energy equivalent to even a small amount of mass is enormous.

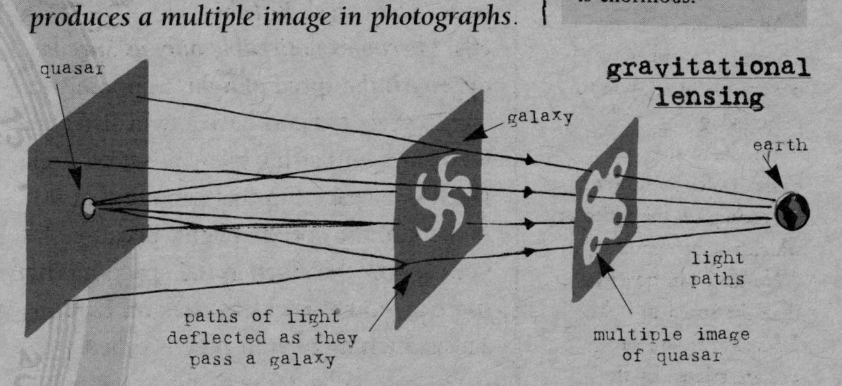

light from the quasar, some 8 billion light-years away, is "bent" by the gravity of a galaxy between the quasar (on the left) and the Earth (on the right)—producing a multiple image, as if viewed through a refracting lens

CHAPTER 2

THE NIGHT SKY

★ The constellations—the name we give to the patterns made by bright stars—help us to find our way around the night sky. They have not changed perceptibly since astronomers first began studying the heavens in earnest some 5,000 years ago, nor are they likely to do so for many millennia to come.

"it looks just like a scorpion!"

Orion's belt

Orion's belt can be clearly seen in the sky

THE CONSTELLATIONS

★ The night sky is full of stars and presents a confusing aspect when you first look at it. But the obvious thing you notice, once your eyes have grown accustomed to the dark, is that some stars are brighter than others. And ***in your mind's eye you can link up the bright stars to form patterns that you can recognize the next time you see them. These patterns made by the bright stars are what we call*** CONSTELLATIONS. They help bring welcome order into our study of the night sky.

> KEY WORDS
>
> **CONSTELLATIONS:** patterns of bright stars that help to map out the night sky

MYTHS AND MONSTERS

★ ***Thousands of years ago, ancient astronomers named the constellations after figures that they imagined the star patterns made. Today we use the names given to the constellations by the ancient Greeks, but in their Latinized form***. The Greeks named the

constellations after the gods and goddesses, heroes and heroines, and animals and monsters that featured in their myths. And they concocted intriguing tales to explain how the figures represented by the constellations ended up in the heavens.

✱ Only a handful of the constellations bear any resemblance to the figures they are meant to portray. For example, it is easy to picture Scorpius with its curved tail of stars as a scorpion poised to sting; and Cygnus does look rather like a swan in flight, with wings outstretched and long neck to the fore. But it's difficult to picture a woman in a chair when looking at the W-shaped pattern of the bright stars of Cassiopeia, or a ram in the inconspicuous stars of Aries.

CARVING UP THE SKY

Astronomers today recognize 88 different constellations (many more than recognized by the ancients). Each constellation refers not just to the bright stars that form the constellation figure, but to **a designated region of the sky and all the stars within it**, with definite boundaries between it and neighboring constellations.

FLYING APART

✱ Because the constellations don't change as time goes by, we might assume that they are fixed in the sky (as the ancients did), or at least traveling through space together as a group. Nothing could be further from the truth. ***With few exceptions, the stars in a constellation pattern have nothing to do with one another and are in reality very far apart. The reason we see them together in a constellation is that they just happen to lie in the same direction in space.***

DOME OF THE HEAVENS

★ When we go stargazing, the night sky seems to form a great dark bowl above our heads, with the stars stuck on it or shining through it. It is the same wherever we go on Earth. In other words, the Earth seems to be at the center of a great heavenly dome.

"it's absolutely heavenly, my dear!"

KEY WORDS

CELESTIAL SPHERE: the spinning dome of the heavens, which has Earth at its center

WHAT YOU SEE

Exactly how you see the stars move in the sky depends on where exactly you are observing from. **At the North and South Poles, the stars will move in circles parallel with the horizon. At the equator, the stars rise and set vertically. At latitudes in between, they rise and set at an angle to the horizon.**

CELESTIAL SPHERE

★ Ancient astronomers thought the sky was a dome and called it the CELESTIAL SPHERE. ***We now know that the celestial sphere is an illusion—there is no dark dome around the Earth with stars stuck on its inner surface. The darkness of the night sky is the void of space, which seems to go on forever, with the stars scattered about within it, all at different distances.***

★ What else did the ancients notice about the night sky? They saw that the stars wheel overhead. ***Rising over the horizon in the east, they arc up through the sky and***

down again as the night goes by, before they disappear over the horizon in the west—so the celestial sphere seems to spin around the Earth once a day, from east to west, carrying the stars with it. Again the ancients got it wrong. It is the Earth that spins around in space, while the stars in the heavens stay still. It spins around on its axis once a day, turning from west to east.

* ***Nevertheless, although the celestial sphere does not exist, the concept is very useful. The stars are so far away that they do appear to be fixed to the inside of a rotating sphere, and astronomers use the geometry of a sphere to pinpoint star positions in the sky*** (***see page 64***).

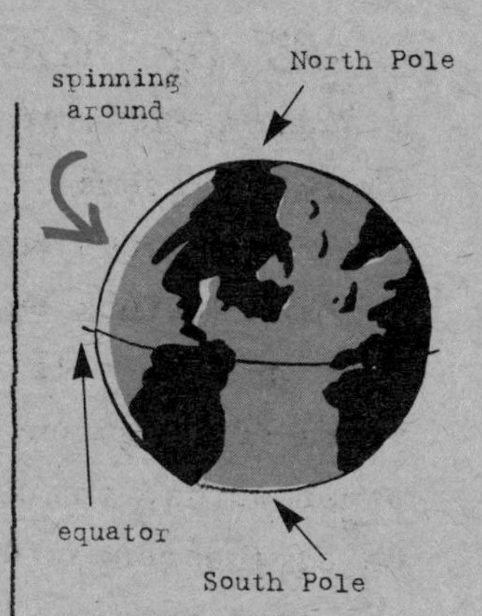

The hemispheres

The celestial equator bisects the celestial sphere. Again by analogy with the Earth globe, we call the two halves the ***northern and southern celestial hemispheres****. Each has its own set of constellations.*

POLE TO POLE

* The Earth spins around an imaginary axis, passing through the North and South Poles. The celestial sphere spins around the same imaginary axis, which touches it at points we call the NORTH AND SOUTH CELESTIAL POLES. ***All the stars in the night sky appear to circle around the celestial poles.***

* Midway between the North and South Poles is the imaginary line around the Earth's circumference that we call the equator. When we apply this idea to the celestial sphere, we get the CELESTIAL EQUATOR. ***In the same way that geographers use the equator for reference on the terrestrial globe, astronomers use the celestial equator for reference on the celestial sphere.***

"must be heading for the celestial pole!"

NORTHERN SKIES

* This star map shows the constellations that occupy the northern half of the celestial sphere. If you are stargazing in the Northern Hemisphere, you will be able to see all of these constellations at some time during the year.

What latitude?

The lines of latitude are invisible lines girdling the Earth. They run parallel to the equator, so your latitude indicates how far north or south of the equator you are.

WHAT YOU CAN SEE

* Which ones you see on any particular night will depend on the time of night and time of year. You will also be able to see some of the constellations of the southern celestial hemisphere. Which of these you can see will depend on the latitude of the place from which you are observing.

* If you live in mid- to high northern latitudes—in northern Europe or North America—you will see some constellations in the northern sky every night as they circle around the north celestial pole (which is marked by THE POLE STAR). These CIRCUMPOLAR CONSTELLATIONS include Ursa Major (Great Bear), Ursa Minor (Little Bear), and Cassiopeia.

what you see depends on how far north or south you are

KEY WORDS

CIRCUMPOLAR STARS: the stars close to the poles, which remain visible all the time

THE MILKY WAY

This hazy white band in the sky is in fact made up of dense concentrations of distant stars. It represents a slice through the plane of our galaxy (see page 101).

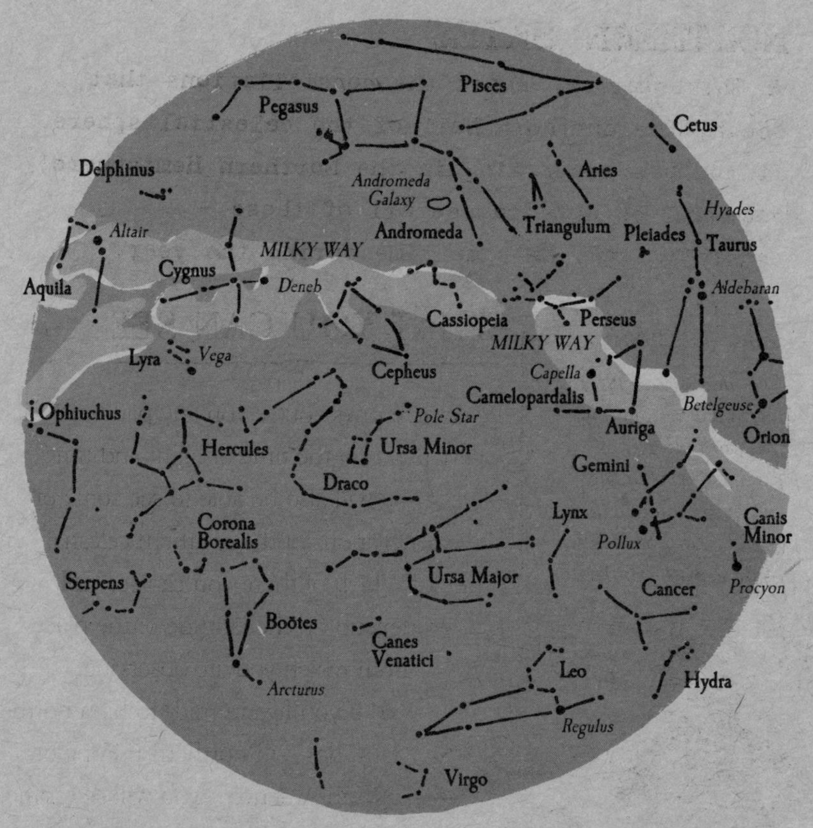

Constellations of the Northern Hemisphere

All the stars appear to circle around the Pole Star, Polaris. Other notable features of the northern skies include the Andromeda nebula and the Pleiades, or Seven Sisters, star cluster.

VIEWING TIME

★ Your time and date of observation will determine which constellations are visible in the night sky. The time is important because different constellations come into view as the celestial dome (or rather, the Earth) spins around (see page 62). The date is important because of the way the Earth travels around the sun during the year (see page 54).

The Lion's shower

Every year, on about November 17, stars seem to rain down from Leo. This is the annual meteor shower known as the Leonids (see page 172).

SOUTHERN SKIES

✱ This star map shows the constellations that occupy the southern half of the celestial sphere. If you are stargazing in the southern hemisphere, you will be able to see all of these constellations at some time during the year.

in mid- to high latitudes you can see the circumpolar constellations

Remember:

Viewing time and date *determine which constellations you can see on a particular night, because the Earth spins around in space once a day like a top and travels around the sun once a year* *(see page 54)**.*

WHAT YOU CAN SEE

✱ Which ones you see on any particular night will depend on the time of night and time of year. You will also be able to see some of the constellations of the northern celestial hemisphere. Which of these you can see will depend on the latitude of the place from which you are observing.

CIRCUMPOLAR STARS

✱ If you live in mid- to high southern latitudes (in Australia, New Zealand, or the southern part of South America or South

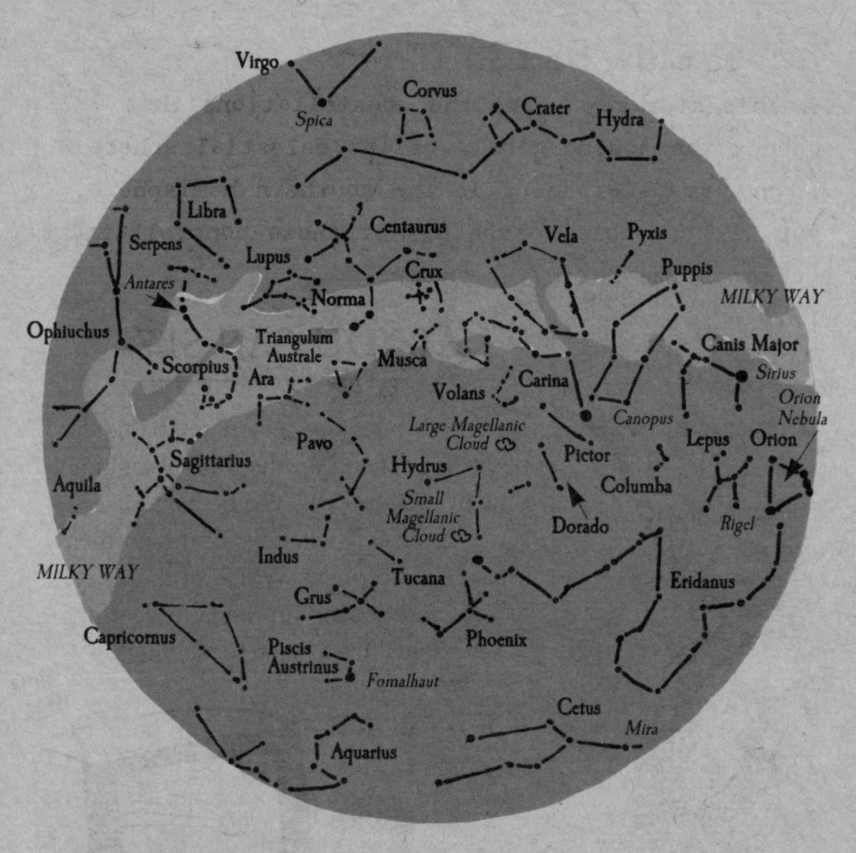

Constellations of the Southern Hemisphere

The constellation Crux points roughly in the direction of the south celestial pole. Lying quite close to the pole are the Large and Small Magellanic Clouds, two small neighboring galaxies.

Africa), you will find that there are some constellations you will be able to see every night in the southern sky as they circle around the south celestial pole. There is no pole star to mark it, however, like there is in the northern celestial hemisphere. These circumpolar constellations include CRUX (THE SOUTHERN CROSS), Centaurus (Centaur), and Carina (Keel).

Remember:

The Milky Way is the hazy white band that arches across the heavens and is a cross-sectional view of the disk of our galaxy (see page 101).

is it July already?

THE CHANGING SEASONS

★ The constellations you see in the night sky change hourly as the Earth spins around. They also change with the seasons. This is a consequence of the Earth making its yearly circumnavigation of the sun.

the stars you see depend on the time and season

it's all changing

KEY WORDS

ECLIPTIC:
the apparent path of the sun through the constellations during the year

VANISHING CONSTELLATIONS

★ If you live at mid-latitudes in Europe or North America, at about 11pm in mid-January, away from city lights, look south and you will see the familiar figure of Orion striding across the heavens. ***Repeat your observation six months later, in mid-July, and Orion is nowhere to be seen. But high above your head will be the three prominent stars of*** THE SUMMER TRIANGLE. Even if you remain stargazing all night, Orion will not appear; and, conversely, in January you won't be able to see the Summer Triangle.

ABRACADABRA

★ ***This celestial vanishing trick has a simple explanation***. Like every other planet, the Earth circles around the sun once a year. From our viewpoint, however, during the course of the year the sun appears to travel around the Earth. During this time it traces a circular path, called THE ECLIPTIC, around the celestial sphere. ★ ***Traveling along the ecliptic, the sun passes through a different constellation each month***. In mid-July the sun is

Orion and the Summer Triangle play seasonal vanishing tricks

traveling through Gemini, so we can't see it because of the sun's glare. In other words, Gemini is always in the sky during daylight, rising and setting with the sun. And so are the constellations around it, including Orion. ***This explains Orion's vanishing trick. In mid-January it appears in our skies during the night, when we are able to see it, but in mid-July it appears during the day, when we can't.*** Similarly, other constellations disappear from the night sky as the sun passes near them, then reappear when the sun has passed by.

FOLLOWING THE CONSTELLATIONS

If you are an experienced stargazer, you can tell the time of year just by checking out which constellations are visible. In the northern hemisphere, as we have seen, Orion dominates winter skies and the Summer Triangle dominates the summer skies. Leo—with its proud head defined by a prominent curve of stars (called the Sickle)—bounds across the heavens in spring. In autumn it is the turn of the Flying Horse, Pegasus, with its distinctive Square. This is also the time of year when we are able to see a misty patch in neighboring Andromeda. Although called the Great Nebula in Andromeda, it is not a nebula but a galaxy. **At a distance of over 2 million light-years, it is the most distant celestial body we can see with the naked eye.**

WINTER AND SUMMER STARS

* Because of Earth's annual journey around the sun, the night sky presents an ever-changing theater of delights for stargazers. Over the next four pages we show contrasting views of the heavens during winter and summer, which will help you familiarize yourself with the heavens.

Orion in love

In Greek mythology, Orion was son of the sea god Poseidon. A mighty hunter, he fell in love with seven beautiful sisters, called the **Pleiades**, *and he still pursues them across the heavens.*

WHICH WAY UP?

* *On these two pages you will find simple star maps that show the constellations in mid-January for the Northern and Southern Hemispheres, at a convenient viewing hour of about 11pm.* At this time of year it is winter in the Northern Hemisphere and summer in the Southern.

* *The view of the night sky for the Northern Hemisphere is looking south. It is in this direction that the greatest changes come about as the seasons go by.* The constellations don't change much looking north because many are circumpolar and

Looking south

some of the stars seen in North America and Europe in January

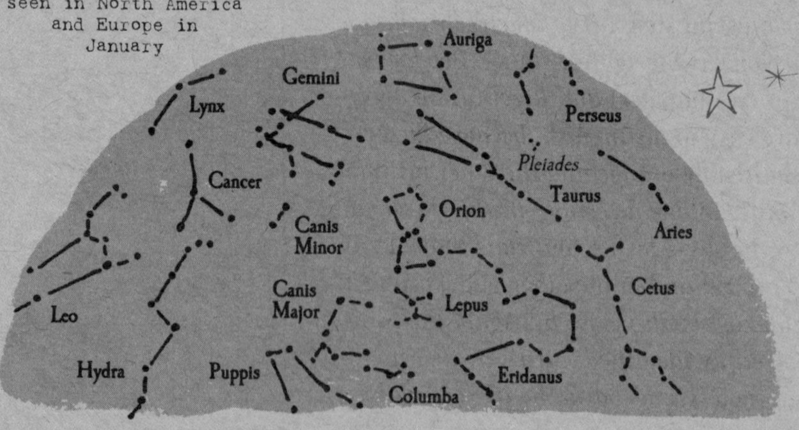

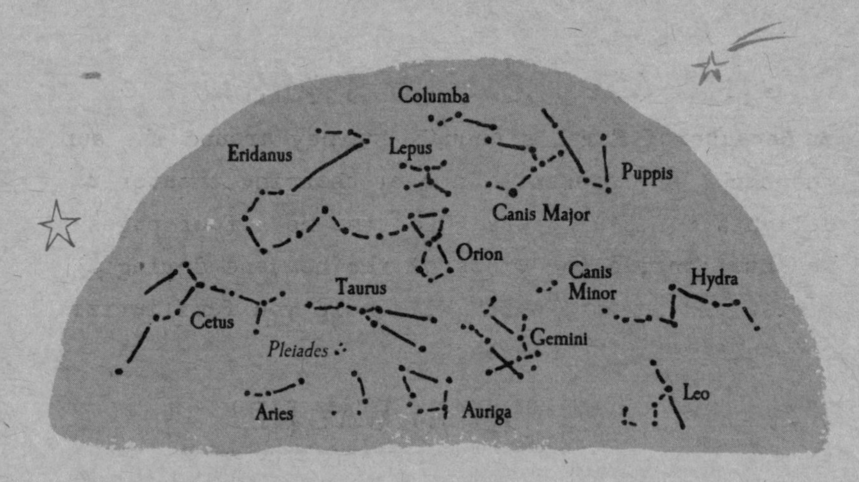

always visible, though they do appear in different positions. ***For similar reasons, the view in the star map for the Southern Hemisphere is looking north, where most changes occur.***

★ ***Viewing at the same time of night, observers in the Southern Hemisphere will see constellations similar to those seen by observers in the Northern Hemisphere but they will appear the other way up.***

Looking north

some of the stars seen in South America and Australia in January

JANUARY SKIES

★ ***These are dominated by Orion, which is a useful signpost to other constellations. Orion's two brightest stars—Betelgeuse (at the shoulder) and Rigel (at the foot)—are easy to distinguish, because the one is noticeably orange and the other brilliant white. In one direction, the brightest star in the sky—Sirius (in Canis Major)—is close by; and in the other, Aldebaran (in Taurus) is also close by. Again the two can be easily identified: Sirius by its brilliant whiteness, Aldebaran by its reddish hue.***

DOG STAR

Brilliant Sirius is also called the **Dog Star** because it is in Canis Major, the Great Dog. It has an unseen companion, which was the first white dwarf star (see pages 79 and 92–3) to be identified.

the Dog Star is the brightest star in the sky

JULY SKIES

★ July skies reveal the Summer Triangle and expose the richest regions of the Milky Way. The sad thing for northern observers, however, is that the summer skies never grow really dark. At this time of the year southern observers are the fortunate ones, because they have dark winter skies.

No accident

As far as the Greek gods were concerned, the absence of Orion from the July skies was deliberate. Because Orion was stung to death by a scorpion, the gods placed the Scorpion in the opposite part of the sky so that it disappears by the time Orion comes into view.

THE SUMMER TRIANGLE

★ In the Northern Hemisphere, the arrival of the Summer Triangle in the night sky signals the arrival (in theory, at least) of hot sunny days and of warm nights for stargazing. ***The Summer Triangle consists of three bright stars—Vega (in Lyra), Altair (in Aquila), and Deneb (in Cygnus). The same three are, of course, also prominent in the Southern Hemisphere, forming a winter triangle.***

★ In the middle of July, in the Northern Hemisphere ***(looking south)*** Altair is in mid-sky at about 11pm, with Vega and

Looking south

some of the stars seen in North America and Europe in July

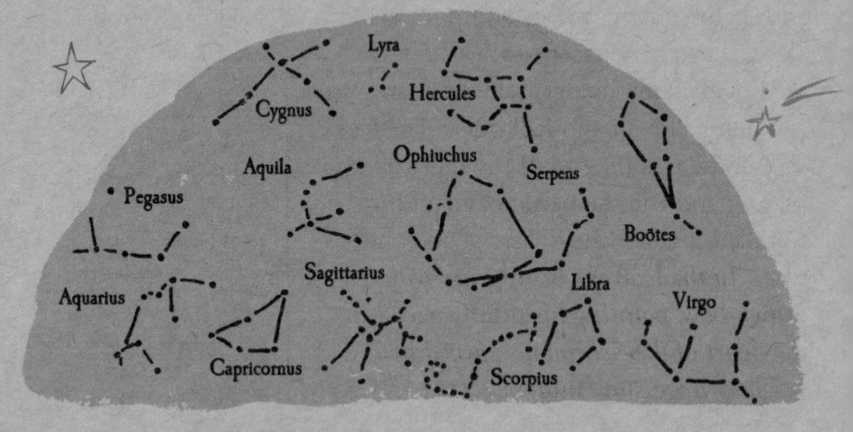

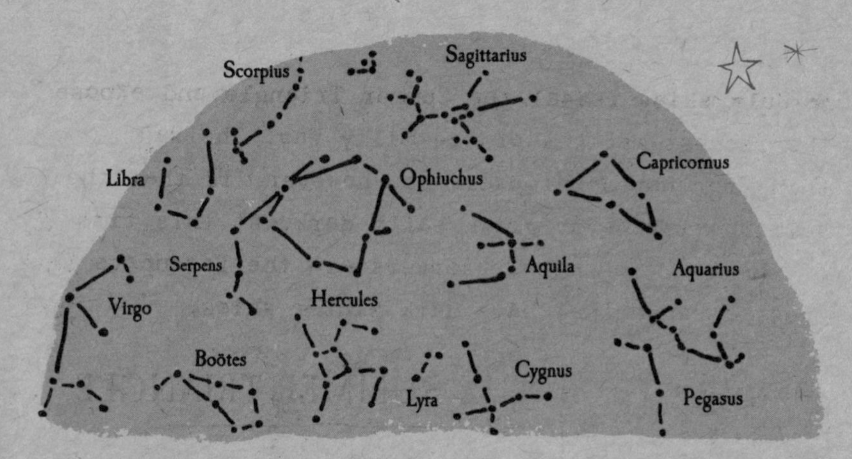

Looking north

some of the stars seen in South America and Australia in July

Deneb visible high overhead. At the same time of night, in the Southern Hemisphere ***(looking north)*** the triangle is inverted, with Vega and Deneb quite low in the sky.

THE ARCHER AND THE SCORPION

★ ***Summer is the time when northern observers can catch a tantalizing glimpse of two of the Southern Hemisphere's finest constellations, Sagittarius (Archer) and Scorpio (Scorpion).*** But unfortunately they appear low down on the horizon where, with light summer nights and polluted skies, they are not seen at their best.

★ In Greek mythology, Sagittarius was a centaur, half man and half horse. Even though he was the son of the musical, pipe-playing god Pan, he became one of the warlike centaurs (there were gentle ones as well). ***In the heavens he appears with a drawn bow, pointing his deadly arrow at the heart of the Scorpion, marked by the bright orange star Antarés.***

RICH MILK

During the northern summer, the Milky Way runs diagonally across the sky from southeast to southwest, through Cygnus, Aquila, Sagittarius, and Scorpius. Even with the naked eye this part of it is exciting to look at, and through binoculars (ideal because of their wide field of view) it is breathtaking. Stars without number light up the sky, and you can see billowing gas clouds, star clusters, and dense globes of stars packed together by the thousand.

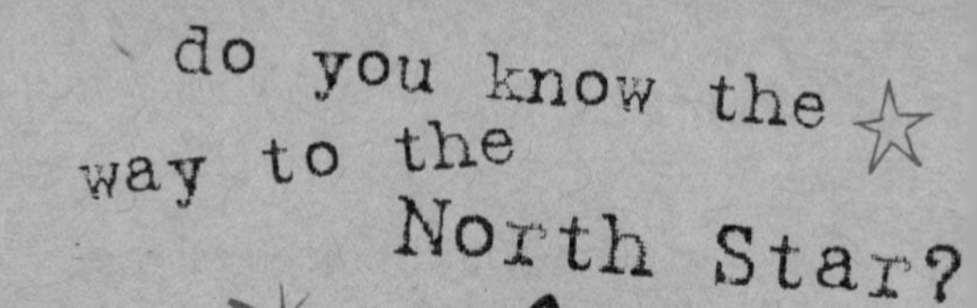

SIGNPOST STARS

★ When you first go stargazing, you have to get your celestial bearings, as it were. The constellations hold the key to this. Many are difficult to make out, but others are easy to find and can, in turn, act as signposts to other constellations and stars.

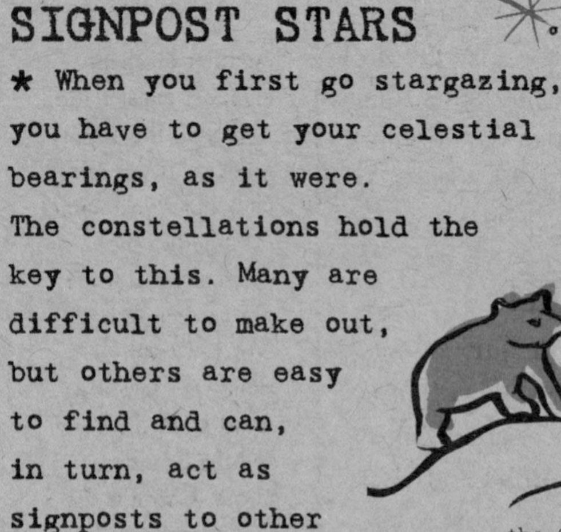

the Great Bear and the Little Bear are next to each other

KEY WORDS

POINTERS: alignments of stars that point to other stars and constellations

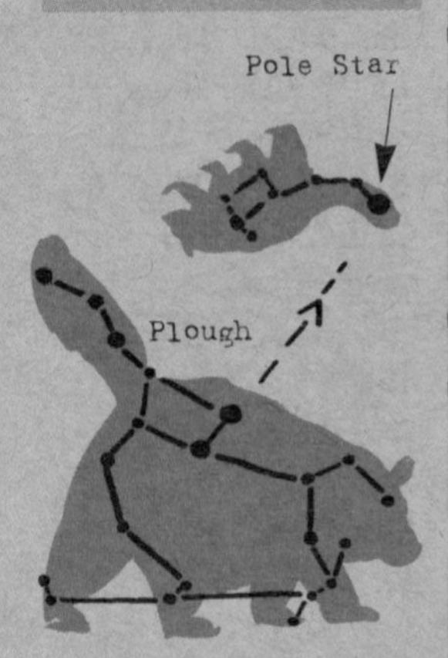

the pointers of the Great Bear point to the Pole Star

POINTING NORTH

★ *In the Northern Hemisphere, there is no finer signpost than the Plow, or Big Dipper, which is part of the constellation Ursa Major (Great Bear). A total of four stars—two at each end (Megrez and Phad, Merak and Dubhe)—define the plowshare.*

★ *Both pairs are useful pointers to other stars and constellations. In fact, Merak and Dubhe are known as* THE POINTERS. A line from Merak through Dubhe points to Polaris (the Pole Star or North Star). This star is part of the fainter constellation Ursa Minor (Little Bear), which looks somewhat like a small plow or dipper. *The Pole Star has been a boon to sailors and navigators for centuries, because when they are facing toward it they know they are looking north.*

★ If you follow a line through Megrez and Phad—the other pair of stars in the plowshare—you come to Leo (Lion) and its bright star Regulus. Other alignments direct you to Gemini (Twins), Auriga (Charioteer), Cassiopeia, Cygnus (Swan), and Boötes (Herdsman).

POINTING SOUTH

★ ***If you go stargazing in the far Southern Hemisphere, the Plow is of no use. You can't see it.*** But there are other excellent pointers, such as Alpha and Beta Centauri—two of the brightest stars in the heavens, which point to the constellation CRUX (THE SOUTHERN CROSS). ***In the Southern Hemisphere there's no pole star as such; but the long axis of the Southern Cross points nearly due south, so southern navigators have no excuse for getting lost.***

★ The Southern Cross also points more or less toward the two misty patches known as THE MAGELLANIC CLOUDS. These are the nearest galaxies to our own.

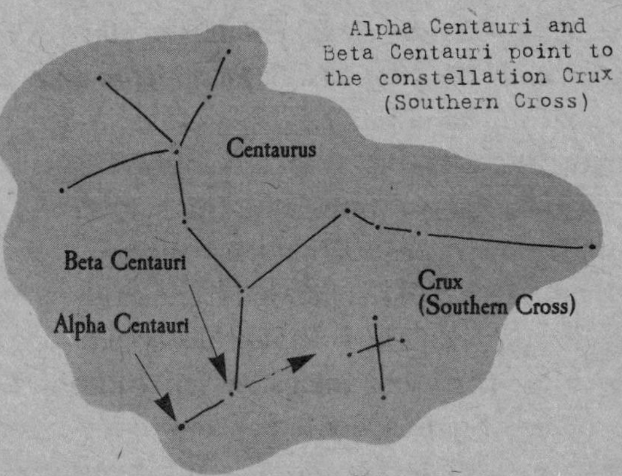

Alpha Centauri and Beta Centauri point to the constellation Crux (Southern Cross)

STRADDLING THE EQUATOR

Another excellent signpost constellation is Orion, the mighty hunter with his raised club and shield. Since he straddles the celestial equator, he can be seen in both Northern and Southern Hemispheres. **Apart from the Plow and the Southern Cross, Orion is the most unmistakable constellation in the heavens. The alignments of the main stars in its distinctive shape point usefully in every direction. The three stars in Orion's Belt point, in one direction, to the brightest star in the sky—Sirius, in Canis Major (Great Dog). In the other, they point to fiery Aldebaran, in Taurus (Bull), and beyond it to the Pleiades cluster.** There are a host of other useful alignments, too, which help make Orion such an invaluable signpost.

STAR TIME

★ **When they are observing, astronomers don't use ordinary time. Their clocks run slightly fast, in order to synchronize with the rotation of the heavens.**

KEY WORDS

SIDEREAL TIME: time relative to the stars (astronomical time)

SOLAR TIME: time relative to the sun (ordinary time)

it's a bit fast today

(to be precise, 23 hours 56 minutes and 4 seconds)

DIFFERENT DAYS

★ The basis of our ordinary measurement of time is the time it takes the Earth to spin around once on its axis. This is the period of time we call the day, and we split it into 24 hours, divided into 60 minutes, which we split into 60 seconds. *But, to be accurate, it is really a solar day and refers to the time it takes for the Earth to spin around once relative to the sun.*

★ *However, this period is not the same as the time it takes for the Earth to spin around once relative to the stars. Because the Earth is traveling around the sun in space, by the time it has revolved on its axis it has moved on a little in its orbit, and relative to the stars has revolved slightly more than once.*

ATOMIC TIME

In modern science, atomic clocks are used to measure time. They do so by counting the unvarying rate of vibration of certain atoms, such as caesium (which vibrates 9,192,631,770 times a second). Such clocks gain or lose less than a second in 2 million years.

for astronomers sidereal time is best, as it matches the timekeeping of the stars

CHECK IT OUT

★ If you time when a particular star rises above the horizon on two successive days, you will find that on the second day it appears 4 minutes earlier. In other words, that star—and, of course, the whole celestial sphere—takes only 23 hours and 56 minutes to spin around once. ***Or putting it the right way around: relative to the stars, the Earth spins around once in 23 hours 56 minutes and (to be precise) 4 seconds.***

SIDEREAL TIME

follow that star

★ ***This time period forms the basis of astronomical time which is referred to as*** SIDEREAL TIME ***(meaning star time).*** It is the sidereal day, and it is made up of 24 sidereal hours, each of which has 60 sidereal minutes, 60 sidereal seconds long. ***Astronomers use sidereal time as a matter of routine, because stars are always in the same position in the heavens at a particular sidereal time.***

Follow that star

To observe a particular star with a telescope, you have to keep moving it to follow the star across the sky as the celestial sphere rotates. *To do this, you first mount the telescope so that it can rotate in the same direction as the stars, on a so-called equatorial mounting (see page 13). Then you can either turn it around very inaccurately by hand, or fit a slow-motion drive that turns around once in every 24 sidereal hours. Because a drive of this kind rotates at the same speed as the celestial sphere, it will keep your telescope locked onto your target star. Such an arrangement is essential for taking long-exposure photographs.*

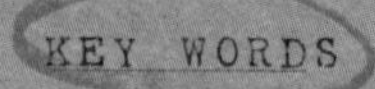

ANGULAR DISTANCE: the distance over a curved surface equivalent to a particular angle at the center of the circle of which the curve is part; the total angular distance over a sphere is 360°

DECLINATION: the celestial equivalent of terrestrial latitude in the system of coordinates for locating a star on the celestial sphere

RIGHT ASCENSION: the celestial equivalent of terrestrial longitude in the system of coordinates for locating a star on the celestial sphere

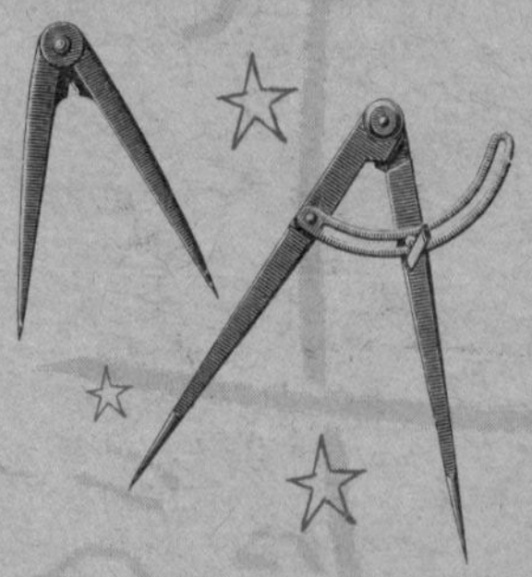

celestial coordinates are used to pinpoint the positions of stars

PINPOINTING STARS

★ **Astronomers use a coordinate system for pinpointing the position of the stars on the celestial sphere, akin to the terrestrial latitude and longitude system used by geographers.**

stargazing is like playing battleships—you need to know the coordinates

GRID REFERENCES

★ *First, let's look at what we do on Earth.* We use a reference grid consisting of two sets of imaginary lines. One set girdles the Earth parallel with the equator (LINES OF LATITUDE), the other spans the Earth from pole to pole (LINES OF LONGITUDE). ***To define the position of a place, we use two*** COORDINATES—measurements, expressed as ANGULAR DISTANCES (degrees), taken from fixed baselines of latitude and longitude.

hello, is that Alpha Centauri?

* For ***latitude***, the baseline is the equator (latitude 0°); from this, latitudes are expressed as so many degrees (0° to 90°) north or south.
* For ***longitude***, the baseline line passes through Greenwich, in London. This is called the prime meridian (longitude 0°). Longitudes are expressed as so many degrees (from 0° to 180°) east or west of the Greenwich meridian.

CELESTIAL LATITUDE

* The DECLINATION (δ), or celestial latitude, of a star is defined in terms of the star's angular distance from the celestial equator. Declinations in the northern celestial hemisphere are given a positive sign (0° to +90°), while those in the southern celestial hemisphere are distinguished by a negative sign (0° to –90°).

CELESTIAL LONGITUDE

* Celestial longitude, or RIGHT ASCENSION (RA), is more complicated. For its baseline, it uses the Point of Aries—one of the points on the celestial sphere where the ecliptic (path of the sun) meets the celestial equator.
* Although right ascension is measured from this point on the celestial equator, it isn't expressed in degrees but in units of sidereal time, ranging from 0 seconds to 24 hours. This isn't surprising, since stars always arrive in the same position in the sky at the same sidereal time.

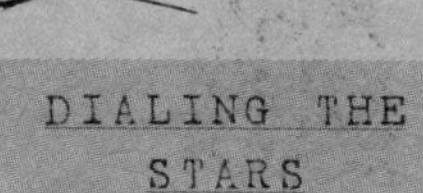

DIALING THE STARS

The concept of **celestial coordinates** is absolutely vital to observational astronomy. By setting right ascension and declination dials on their telescopes to the coordinates of the star they want to study, astronomers can get it in their sights immediately.

the lines of latitude and longitude form a grid around the globe

"The fault is in our stars..."

THE SUN AND STARS

★ The sun circles around the celestial sphere every year, and it passes through a different constellation every month. These constellations have always been considered special by astrologers, who regard them as influencing human affairs.

THE CIRCLE OF ANIMALS

★ *The sun travels along the ecliptic during the year. The planets are never found far away from it either*. The reason is that the sun, the Earth, and the planets all travel through space in much the same plane (see page 113). ***As a result, the planets are always found within an imaginary band in the heavens, on either side of the ecliptic. We call this band*** THE ZODIAC, ***and the stars it passes through are known as the*** CONSTELLATIONS OF THE ZODIAC. The term zodiac means "circle of animals"—because most of the constellations that the sun passes through have animal names.

ONE STAR SIGN SHORT

★ ***Most astronomers regard astrology as having no validity.*** As scientists, they can find no shred of evidence to support astrologers' claims and can point out a

KEY WORDS

ZODIAC: an imaginary band in the heavens through which the sun and planets appear to travel

PRECESSION: slight wobbling of the Earth as it spins on its axis (something like the wobbling of a spinning top)

"I might have guessed she was a Taurean!"

number of things that don't add up. ***For example, the sun actually passes through 13 constellations during the year. The extra one not included in the astrological zodiac is Ophiuchus, the Serpent-bearer.*** The sun passes through it in November on its way from Scorpius (always called Scorpio by astrologers) to Sagittarius and spends far more time there than it does in Scorpius. ***"And what about*** PRECESSION***?" they also cry.***

THE CONSTELLATIONS OF THE ZODIAC

In the classic zodiac beloved of astrologers there are 12 constellations, which they call **star signs.** Astrologers divide up the year into 12 time periods, assigning a star sign to each, and assume that the sun passes through the signs during these periods. According to astrology, the constellation in which the sun happens to be at the time of a person's birth determines how the heavenly bodies will affect that person's life.

BIRTHDAY	STAR SIGN
Mar 21–Apr 20	Aries (Ram)
Apr 21–May 21	Taurus (Bull)
May 22–June 21	Gemini (Twins)
June 22–July 23	Cancer (Crab)
July 24–Aug 23	Leo (Lion)
Aug 24–Sept 23	Virgo (Virgin)
Sept 24–Oct 23	Libra (Scales)
Oct 24–Nov 22	Scorpio (Scorpion)
Nov 23–Dec 21	Sagittarius (Archer)
Dec 22–Jan 20	Capricorn (Sea Goat)
Jan 21–Feb 19	Aquarius (Water-bearer)
Feb 20–Mar 20	Pisces (Fishes)

What about precession?

The dates for the star signs astrologers use, which supposedly relate to the period when the sun is in a particular constellation, hark back to the zodiac of Roman times, some 2,000 years ago. But things have moved on since then. Because of a gradual wobbling in the Earth's axis, known as precession, the apparent path of the sun through the stars has changed. In Roman times the sun crossed over the celestial equator when traveling through Aries. Today it is traveling through Pisces when it crosses. So all the zodiacal constellations used by astrologers are now one sign out. Game, set, and match to astronomers?

Friedrich Bessel

CHAPTER 3

THE STARS

★ The reason why the stars look like they are fixed in their constellations is not because they don't move through space, but because they lie such incredible distances away. Even the light from nearby stars takes years to reach us, and light from the most distant objects we can see takes billions of years.

PARALLAX:
the principle that a nearby object appears to move against its background, when it is viewed from different places

LIGHT-YEAR:
the distance light travels in a year

BESSEL'S BOMBSHELL

★ Even in the most powerful telescopes, the stars are never more than pinpoints of light. So they must lie very far away. No one had any idea of just how far until 1838. *In that year German astronomer* FRIEDRICH BESSEL *(1784–1846) made the first accurate calculation of the distance to a star—the relatively nearby star 61 Cygni, in the constellation Cygnus. It proved to be an astonishing 65.2 million million miles away.*

★ *At this distance, its light takes 11 years to reach the Earth, so astronomers say that it lies*

I make it about 11 light-years

65,000,000,000,000 miles as the swan flies...

11 light-years away. They use the LIGHT-YEAR as a convenient yardstick for measuring astronomical distances. ***The farthest objects we have detected in the universe, quasars, appear to lie as much as 13 billion light-years away (in miles, this would be something like 7 followed by 22 zeros).***

you must be **joking**

even the nearest stars are millions of miles away

PARALLAX

★ Bessel used the method of PARALLAX in his calculations, a useful method for calculating the distance to nearby stars. The principle of parallax is simple. Hold up a finger in front of your face, and look at it first with one eye, then with the other. You notice that your finger appears to move against a distant background. ***In a similar way, if you look at a nearby star from two different vantage points, it appears to shift against a background of more distant stars.***

★ ***For maximum shift, astronomers choose the most widely separated vantage points possible—they view the star from opposite ends of the Earth's orbit around the sun at six-monthly intervals. From the extent of the star's shift in position and knowing the distance between the two vantage points (the diameter of the Earth's orbit), they can work out the distance to the star by simple trigonometry.***

PARSEC

Astronomers use the parallax principle to define another unit of distance, **the parsec.** It is the distance at which a star would show a parallax of 1 second of arc, or 1/3600 of a degree. **1 parsec is equivalent to about 3.3 light-years.**

THE LIGHT-YEAR

Light travels at a speed of 186,390 miles per second. In one year, it travels about 5.9 million million miles.

BRIGHT SHINERS

* Even a casual glance at the night sky shows that stars vary greatly in brightness. While some are scarcely visible, others stand out like beacons. Astronomers tell us that, in absolute terms, some stars shine millions of times brighter than the sun.

Hipparchus

KEY WORDS

ABSOLUTE MAGNITUDE:
a measure of the true brightness of a star

APPARENT MAGNITUDE:
a measure of how bright a star appears to be from Earth

LUMINOSITY:
the true brightness of a star, reflecting its total energy output

STAR MAGNITUDE

* Astronomers express the brightness of a star by its MAGNITUDE, following the system devised by the Greek astronomer HIPPARCHUS in about 150 BC. *He classed the brightest stars visible to the naked eye as stars of the 1st magnitude, and ones just visible as stars of the 6th magnitude.*

* Modern light-measuring instruments have brought precision into the system. *A star of the 1st magnitude is 100 times brighter than one of the 6th magnitude. Magnitudes are now expressed to one or two decimal places as well. In addition, a few stars in the sky are noticeably brighter than the average 1st-magnitude stars, so have been given* NEGATIVE MAGNITUDES. Sirius, the brightest star in all of the heavens, has a magnitude of –1.45.

★ *In telescopes, of course, stars that are invisible to the eye become visible. And so the magnitude scale has extended beyond 6 to higher values. Modern telescopes can detect objects of the 25th magnitude—millions of times dimmer than visible stars.*

ABSOLUTE ACCURACY

★ *All the stars lie at different distances from us, so a truly bright star a long way away could appear fainter than a truly dim star much closer to us.* The brightness of a star as seen from the Earth is therefore termed its APPARENT MAGNITUDE.

★ To provide meaningful comparisons between the true brightness of stars, we would have to view them from the same distance. This is the basis of the ABSOLUTE MAGNITUDE scale, *which indicates the magnitude that would be observed if the star was at a distance of 10 parsecs (33 light-years).*

★ Absolute magnitudes immediately put things into perspective. *Sirius, which to the eye seems the brightest star in the sky, has an absolute magnitude of about +1.4. Yet the apparently dimmer star Rigel, in Orion, is really very much brighter, with an absolute magnitude of –7. It outshines our dim sun (+4.8) by as much as 50,000 times.*

BRIGHTNESS AND DISTANCE

Calculating the distance to stars is a pain in the neck because they lie so far away. But if we know a star's true brightness, we can easily calculate how far away it is by comparing the true (absolute) brightness with the brightness we see. This is because a well-known law of physics **(the inverse-square law)** tells us that doubling the distance reduces the brightness by 2^2, or 4 times. This method enables us to estimate distances to Cepheid variable stars (see page 84–5), because their observed light variation tells us their absolute brightness.

STARLIGHT MESSAGES

* The faint light coming from distant stars is supercharged with information. Astronomers extract this information via the little smudge of color they get when they pass the light through a spectroscope.

THE STELLAR RAINBOW

* *Light from the stars is nominally white, just as is sunlight.* But if we pass this white light through a spectroscope, we get a band of color—THE SPECTRUM. *This shows the same rainbow spread of colors we see in the spectrum of sunlight, from blue to red. The color is our visual perception of the wavelengths of light—shortest for blue light, longest for red.*

"That's not a stellar spectrum! Where are the dark lines?"

SPECTRAL SIGNATURES

The dark lines appear in a star's spectrum as a result of certain wavelengths of light being absorbed by chemical elements present in the star's cooler atmosphere. Each chemical element has a characteristic set of lines—its "spectral signature"—that enables us to tell which element is present. The nature and significance of the spectral lines in starlight were first elaborated by **Robert Bunsen** (1811–99), inventor of the Bunsen burner, and **Gustav Kirchoff** (1824–87). This followed earlier investigation of the dark lines in the solar spectrum by **Josef von Fraunhofer** (see page 120).

★ The spectrum varies from one star to another in many ways. For example, ***some stars are brightest in the blue part of the spectrum, some in the yellow, and some in the red. This tinges their light blue, yellow, and red respectively. We can often see this color difference in the night sky***. Rigel, in Orion, is definitely bluish-white. Aldebaran, in Taurus, is distinctly reddish. Our own sun is definitely yellowish.

★ ***This variation in brightness within the spectrum is related to the surface temperature of the stars. Bluish stars are very hot (36,000°F or more), yellowish stars medium hot (11,000°F), and reddish stars cool (5,500°F).***

TELL-TALE LINES

★ ***Close observation of the*** STELLAR SPECTRUM ***reveals that it isn't continuous. Cutting across it at intervals, all the way along, are numerous dark lines. More of these*** SPECTRAL LINES ***appear in the spectra of cool stars than in hot ones.***

★ ***Spectral lines furnish astronomers with a great deal of information. For example, because every substance has a unique*** SPECTRAL SIGNATURE, ***they reveal a star's composition***. Hydrogen and helium make up over 90 percent of ordinary stars, and so they show up strongly. But astronomers also commonly find iron, nickel, carbon, silicon, sodium, calcium, nitrogen, and oxygen. Altogether, about 70 different elements are found.

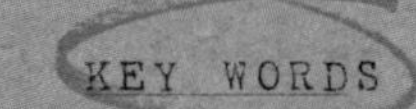

SPECTRAL LINES: dark lines in the spectrum of starlight

"Add some iron and a pinch of oxygen..."

Magnetic splits

In the spectrum of some stars, the dark lines are split due to the ***Zeeman effect****, which happens if the star has a very intense magnetic field. Some stars' magnetism is many thousand times stronger than the sun's. If we lived on a planet orbiting around a star like this, we could generate electricity for our homes by rigging up coils of copper wire on the walls. The rotation of the planet in the star's powerful magnetic field would make electric current flow in the coils.*

THE SHIFTING LINES

★ It isn't just the nature of the spectral signatures that astronomers find meaningful, it is also their position within the spectrum. A shift in the lines, this way or that, tells us how fast a star is moving and in which direction. The shift in the spectral lines of galaxies reveals the nature of the universe.

KEY WORDS

Blue shift: the shift of spectral lines toward the blue end of a star's spectrum

Red shift: the shift of spectral lines toward the red end of a star's spectrum

Doppler effect: the apparent change in the wavelength of light or sound waves when the source of the waves is moving

BLUE OR RED?

★ The sun—a yellow star with a surface temperature of about 10,000°F—produces a dark-line spectrum in which the tell-tale lines occupy certain positions. *But in the spectra of other stars, the lines are often in different positions. They may be shifted toward the blue end of the spectrum (blue shift), or toward the red end (red shift).*

★ ***Astronomers interpret these shifts as a Doppler effect (see right), caused by the star's motion***. If the star is moving toward us, the light waves bunch up in front of it, so they appear to have a shorter wavelength and are therefore bluer. If the star is moving away from us, the waves are stretched out, so they appear to have a longer wavelength and are therefore redder. ***The amount of the red or blue shift is a direct measure of how fast the star is moving toward or away from us.***

The Doppler effect

We often experience the Doppler effect with sound waves. If a police car races past with siren screaming, we hear a high-pitched note (short wavelength) as it approaches and a much lower-pitched note (longer wavelength) as it recedes.

RED SHIFTS

★ On average, it appears that there are about as many stars moving toward us as there are traveling away from us. This is what we might expect in view of the fact that we are located within the interior of a rotating galaxy (see page 101).

★ ***When we examine the spectra of other galaxies, however, we find that they almost all show a red shift. They must therefore be rushing away from us. The extent of the red shift gives the speed at which the galaxies are receding. The interesting thing is that galaxies move faster the farther away they are, so the extent of their red shift provides a measure of their distance from us***. The application of this principle has placed some of those enigmatic bodies we call quasars at the edge of the observable universe, with a speed of recession approaching the ultimate—that of light.

VESTO SLIPHER (1875-1969)

US astronomer **Vesto Slipher** first discovered the red shift of certain galaxies in 1912. But the significance was not appreciated, since they were then thought to be nebulae that formed part of our own galaxy. Later, **Edwin Hubble** discovered their true nature and related their headlong flight through space to the concept of an expanding universe.

A GOOD PLOT

* Every scientist loves a good graph to show relationships in the data that has been acquired. Astronomers are no exception, and they come up with an excellent example when they plot the spectral class of stars against their luminosity, or true brightness.

good scientists pride themselves on their graphs

easily committed to memory and politically incorrect

KISS ME RIGHT NOW!

* The stellar spectrum is an excellent way of classifying a star. *There are 11 main spectral classes—W, O, B, A, F, G, K, M, R, N, and S. These can be memorized using the politically incorrect mnemonic: "Wow! Oh Be A Fine Girl, Kiss Me Right Now, Sweetie!" This sequence ranges from very hot (W) to very cool (S). Each class is divided into 10 subclasses, from 0 to 9.*

THE H–R DIAGRAM

* And now we come to the plot. In the early 1900s, the Danish astronomer EJNAR HERTZSPRUNG (1873–1967) and the US astronomer HENRY RUSSELL (1897–1957) independently began relating spectral class to the LUMINOSITY—or absolute brightness—of stars. The result was the HERTZSPRUNG–RUSSELL (H–R) DIAGRAM, in which a star's luminosity, or absolute magnitude, is plotted on a

y-axis (vertical) against spectral class, or temperature, on the x-axis (horizontal).

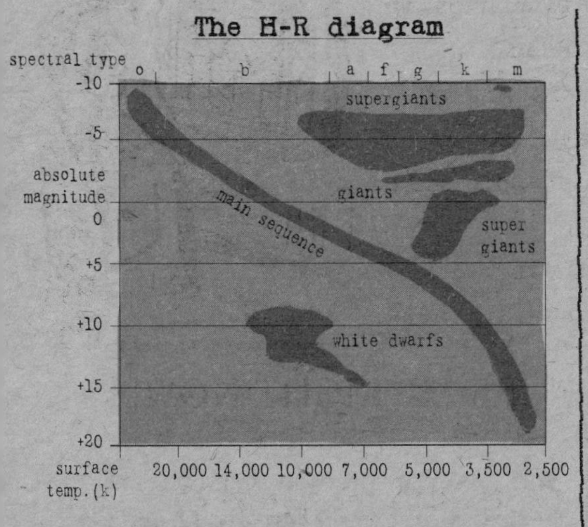

THE MAIN SEQUENCE

★ *In the H–R diagram, the majority of stars lie along a roughly diagonal band, known as the* MAIN SEQUENCE. *Stars that appear on this band are in the prime of their life, and shine steadily. They leave the main sequence when they start to die.*

★ The position of a star on the band and how long it remains there depend chiefly on its mass. *M-class stars near the bottom of the graph are small, cool, and red, and shine steadily for 100 billion years or more. G-class stars, like the sun, appear about halfway up: they are medium-sized, medium hot, and they shine for about 10 billion years. B-class stars near the top are very large, very hot, and bluish. They lead brief showy lives, remaining on the main sequence for mere millions of years.*

GIANTS AND DWARFS

When stars cease shining steadily and start to die (see pages 92–3), they leave the main sequence, drifting to the right on the H–R diagram. That is where we find the bigger and brighter, but cooler and redder, giant and supergiant stars like Antares and Betelgeuse. The supergiants inevitably explode and blast themselves apart, and in so doing disappear off the diagram. The giants, however, eventually shrink and dim, but at the same time heat up. Consequently they cross back over the main sequence and end up beneath it as white dwarfs (see pages 92–3), like the original and best-known one—the companion of Sirius, known as Sirius B.

are you a white dwarf?

STELLAR PROFILE

★ So what is a typical star? The truth is, there isn't one. As we have seen on the previous pages, they vary wildly in size, mass, color, brightness, temperature, speed, distance, age, and so on. If we had to choose a typical star, it would be the one that is on our own cosmic doorstep —the star we call the sun.

Primeval hydrogen

Hydrogen is the simplest, lightest, and most common element in the universe, formed shortly after its beginning nearly 5 billion years ago.

what is a typical star?

MAIN-SEQUENCE STARS

★ The reason for choosing the sun is that it is a stable main-sequence star. ***So let's sum up the characteristics of a stable main-sequence star.***

★ It is a huge globe of hot gas, mainly hydrogen and helium.

★ It produces the energy that keeps it shining in its interior by nuclear-fusion reactions, in which hydrogen acts as fuel.

★ It pours out its energy into space as visible light (ranging from blue to red) and also as invisible rays, ranging from X-rays to radio waves.

★ In size, it can vary from about one-twelfth the diameter of the sun to 50 or more times the sun's diameter.

* The surface temperature may be as low as 3,500°F (for a tiny red star) or as high as 63,000°F (for a big blue-white star).
* It can be a dim candle of a star, with only about one-hundredth the brightness of the sun, or a blazing beacon that is tens of thousands of times brighter than the sun.

GIANT STARS

* ***The giant and supergiant stars that we find off the main sequence are not as stable. They are usually changing quite rapidly, cosmically speaking, and are nearing the end of their lives (see page 94). Here are some of the most important ways in which giants and supergiants differ from main-sequence stars.***
* The nuclear reactions taking place inside them are different because giants and supergiants have used up most of their hydrogen.
* In diameter, giants can be up to about 100 times larger than the sun—and supergiants as much as 500 times larger.
* Giants and supergiants are all cool and red.
* Because of their enormous size, however, they are all very luminous—and in some cases hundreds of thousands of times brighter than the sun.

White dwarfs

A white dwarf represents a late stage in the life story of a star like the sun. Here is how it differs from a main-sequence star.
** No nuclear reactions are taking place inside it.*
** The energy that keeps it shining has come from the gravitational collapse of its matter.*
** It is tiny—about the size of the Earth or smaller.*

** Typically, it has the mass of the sun—and so enormous density. A teaspoonful of its matter would weigh 5 tons.*
** Its temperature can be 27,000°F or higher.*
** But, being small, its luminosity is low; usually about a thousandth of the luminosity of the sun.*

SEEING DOUBLE, AND TRIPLE...

* Most stars do not travel through space alone, like the sun does. The majority have one or more traveling companions, which interact with one another.

KEY WORDS

BINARY: a double-star system in which the stars circle each other

ECLIPSING BINARY: a binary in which the two stars periodically eclipse each other, leading to a variation in brightness

Good companions

Out of every 100 stars, nearly half are binaries and more than a fifth have three or more components. Only about 30 travel through space alone.

MARVELOUS MIZAR

* For stargazers in the Northern Hemisphere, the star pattern we call the Plow or Big Dipper is a familiar friend. *If you have reasonable eyesight, you will see that Mizar—the second star in the handle of the Plow or Dipper—has a fainter sidekick, Alcor. We call such a pair of stars a double star. But Mizar and Alcor are not traveling together. In fact they are about 10 light-years apart, and just happen to lie in the same direction in space.*

* However, look at Mizar in a telescope and you will find that it is not one but two

stars. And this time the two stars do travel together. They orbit around each other, or rather around a common center of gravity (THE BARYCENTER). We call this kind of double-star system a BINARY—and if its two components are visible through a telescope, it is known as a VISUAL BINARY.

"Come this way to see the Winking Demon."

LOOK AT THE LINES

* But we haven't finished with Mizar yet. When we examine the light spectrum of each of its two components, we see periodic shifts in the spectral lines (see pages 74–5). This tells us that each component is itself double, ***the shift in the lines reflecting the movement of the unseen components toward or away from us as they orbit around each other***. This kind of binary, only detectable in its spectral lines, is called a SPECTROSCOPIC BINARY. ***So Mizar is in fact a quadruple star with an apparent hanger-on, Alcor.***

THE WINKING DEMON

The Arabs gave the second brightest star in the constellation Perseus the name Algol—meaning the demon star, or winking demon. The reason is that every 69 hours it dims from second to third magnitude, over a period of about four hours. It then takes about the same time to regain its former brilliance. A young English astronomer, **John Goodricke** (1764–86), was the first to explain why Algol behaves in this way. **It is, in fact, a binary star system composed of a large dim star and a small bright one. The two stars orbit in our line of sight, so that the dim star periodically passes in front of the bright one, eclipsing it and causing the overall brightness of the system to diminish.** We now know of many other **eclipsing binaries** like this.

the Jewel Box (in the Southern Cross) got its name because its stars flash many colors

CLUSTERING TOGETHER

★ Although many stars travel through space with one or more close companions, some travel together in great clusters containing hundreds or even hundreds of thousands of stars. Some of them we can see with the naked eye.

Southern gem

One of the loveliest open clusters is visible only in the Southern Hemisphere, in the Southern Cross. Because its stars flash many colors, John Herschel (son of William Herschel, the discoverer of Uranus) named it the Jewel Box.

SEVEN SISTERS, OR MORE

★ *In the Northern Hemisphere, the easiest star cluster to see is in the constellation Taurus (the Bull)*. It lies not far away from the orange star Aldebaran, which marks the Bull's eye. If the night is clear and you have good eyesight, you may be able to make out the six brightest stars in the cluster, although in the past seven have been observed, hence the common name for this cluster—THE SEVEN SISTERS.

amorous Orion still pursues the Seven Sisters

★ *The Pleiades, to give it its proper name, is an example of what astronomers call an* OPEN CLUSTER, *in which the stars are relatively far apart. Telescopes show that it contains not just seven stars, but probably as many*

as 300. As in other open clusters, the stars are young—only about 50 million years old. They are hot blue stars surrounded by nebulosity (hazy dust clouds).

* There is also another open cluster in Taurus, scattered around the Bull's eye. ***Called the Hyades, at a distance of some 140 light-years it is the open cluster nearest to us***. About 200 stars belong to this cluster (Aldebaran isn't one of them).

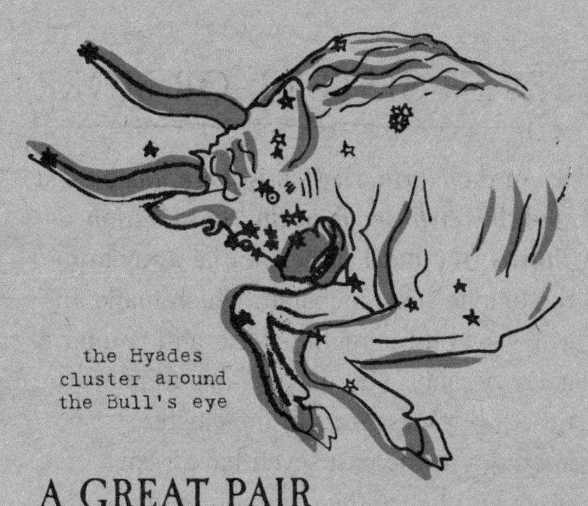

the Hyades cluster around the Bull's eye

A GREAT PAIR

* Two of the most conspicuous clusters in the Southern Hemisphere, however, are not of the open variety. In fact they couldn't be more different. The two clusters, which look like particularly prominent stars and have star names—Omega Centauri and 47 Tucanae—are termed GLOBULAR CLUSTERS. This is because they have their stars packed closely together into a globe shape. ***These clusters don't just have a few hundred stars—they have hundreds of thousands***.

KEY WORDS

GLOBULAR CLUSTER: a globe-shaped concentration of hundreds of thousands of stars

OPEN CLUSTER: a group of several hundred stars traveling through space together

Globular clusters

Globulars differ from open clusters in a number of ways. They are not found in the disk of our galaxy, as are open clusters and ordinary stars. Nor do they take part in the general whirlpool-like circulation of the galaxy. Instead, they circle independently around the galaxy's central bulge, at a great distance from it. Furthermore, while the stars in open clusters are very young, those in globular clusters are very old. In fact, astronomers believe that they were probably formed at the same time as the galaxy itself.

I'll just turn myself down

VARIABLE STARS

* Most stars shine steadily, year in, year out. True, they twinkle in the night sky—but that is an atmospheric aberration and has nothing to do with the stars themselves. Other stars, however, vary noticeably in brightness from time to time because of processes going on within them. And some of them vary with clockwork precision.

"But it was there yesterday!"

KEY WORDS

VARIABLES: stars that vary in brightness

MIRA STARS: variables that change in brightness over a long period

CEPHEIDS: variables that vary in brightness with absolute precision in a relatively short period of time

MIRA THE WONDERFUL

* In 1596 a Dutch astronomer named DAVID FABRICIUS (1564–1617) noted how a star in the constellation Cetus began fading and within a few weeks disappeared. ***This was the first recorded sighting of a variable star. The one that he discovered came to be called Mira, meaning "the wonderful." Many stars like Mira have since been found and are, appropriately, named*** MIRA STARS.

* Mira itself varies quite dramatically. Its brightness peaks at 2nd magnitude, and it is then easily visible to the naked eye. But within about 11 months it has faded to 10th magnitude, and ceases to be visible even with binoculars. Other ***Mira stars fade and brighten over different periods, ranging from about 2½ to 20 months.***

All Mira stars are red giants. Astronomers reckon that they vary in brightness because they physically pulsate, alternately expanding in size and becoming brighter then shrinking and fading.

CLOCKWORK CEPHEIDS

* ***Mira stars are not regular in their habits—their period of variation alters, as do their highs and lows of brightness.***
* ***But in the constellation Cepheus there's another type of variable star, Delta Cephei, which brightens and dims with absolute regularity.*** A pulsating yellow giant, it varies in brightness between magnitudes 3.5 and 4.3 in precisely 5.366 days. ***This star gives its name to the*** CEPHEIDS, ***a class of pulsating variables with habits as regular as clockwork***. Among them is Polaris, the Pole Star, although its dip in brightness, over a period of almost four days, is scarcely perceptible to the naked eye.
* ***Typical Cepheids have a period ranging from about 1 to 50 days.*** RR LYRAE VARIABLES ***are closely related, but have much shorter periods, measured only in hours***. These are also much older stars. Often called CLUSTER VARIABLES, they are to be found in the globular clusters that surround the centers of galaxies.

Henrietta's law

*US astronomer **Henrietta Leavitt** (1868–1921) became noted for her work on Cepheid variables at Harvard Observatory. In 1912 she found that the period of variation of a Cepheid is directly related to its absolute (true) brightness, or luminosity. This discovery gave astronomers a valuable yardstick for estimating distances to remote galaxies* *(see page 71).*

ERUPTING VARIABLES

★ While the Cepheids and their pulsating kin vary in brightness predictably, other kinds of variable stars behave erratically. They suddenly flare up without warning, and in some instances may blast themselves apart in a catastrophic explosion.

KEY WORDS

FLARE STAR: a red dwarf star that brightens periodically

NOVA: a star in a binary star system that suddenly explodes and flares up dramatically

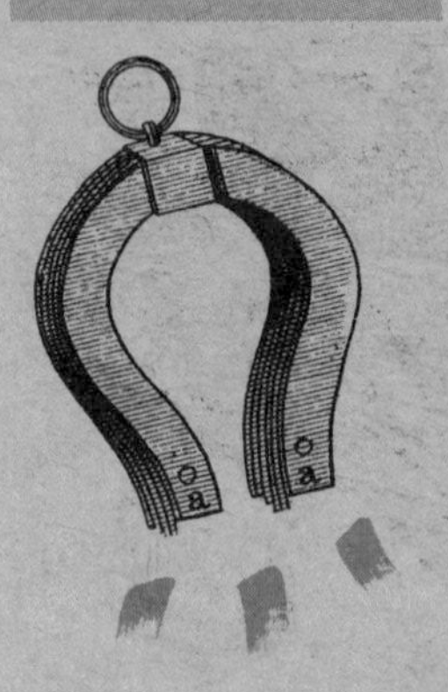

flares may be caused by tortuous magnetic fields

FLARING UP

★ Variable stars that suddenly flare up are called ERUPTIVE VARIABLES. *Among them are some dwarf red stars known as* FLARE STARS. *Almost daily they brighten up for a few minutes when they emit flares of incandescent gas similar to those we see coming off the sun*. Flares are thought to be caused by tortuous magnetic fields releasing energy. On the sun, the brightness of the flares is lost against the general surface glare. But red dwarfs are so dull that the flares brighten up the whole star.

A NEW STAR THAT ISN'T

★ *From time to time, what seem to be new stars suddenly appear in the night sky, shine brightly for a while, and then fade away. Although they are known as* NOVAE *(meaning "new"), through a telescope such "new stars" are revealed to be existing stars that have suddenly brightened dramatically. As many as 30 may occur in our galaxy in the course of a year.*

★ *Astronomers think that novae occur in close binary star systems that include a white dwarf and a red giant. Gas streaming from the giant to the dwarf builds up and heats up, until eventually it becomes hot enough for nuclear fusion to take place. As fusion ripples through the piled-up matter, the star starts to brighten. Within a few hours, or a few days at most, it becomes a new stellar beacon—up to a million times brighter than it was originally. It carries on shining until all the accumulated matter has gone, then fades back to normal. This may take a few years or just a few months, as with the much studied Nova Cygni (in the constellation Cygnus) in 1992.* The Hubble Space Telescope was launched in time to follow the expanding shell of gases blasted into space during the nova outburst.

★ *Sometimes the nuclear process is violent enough to blast the star apart. We then see an even more brilliant flare-up, called a* SUPERNOVA—*or, to be more precise, a Type I supernova. This is more brilliant than a Type II supernova, which occurs when a massive star dies* *(see page 95).*

SOOTY STAR

Another kind of variable is typified by the star R Coronae Borealis. A supergiant with an atmosphere rich in carbon, it shines steadily for most of the time, but every now and then it suddenly dims dramatically. Astronomers reckon this happens when it puffs off clouds of carbon particles (in other words, soot), which block its light.

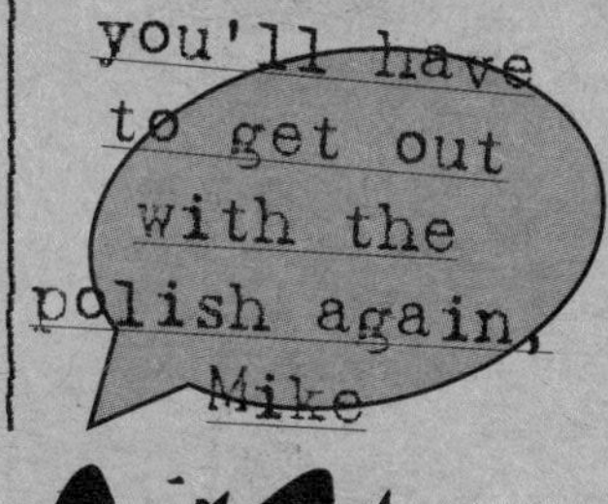

BETWEEN THE STARS

★ Most of the matter in the universe is found concentrated in the stars, but the space between the stars is not completely empty. Scattered around in minute quantities almost everywhere, there are wisps of gas and specks of dust. In places, this interstellar matter gets denser and forms great clouds many light-years across.

"That interstellar matter is so dense!"

COSMIC CLOUDS

★ Astronomers call the billowing clouds of gas and dust NEBULAE, the Latin word for clouds. *The nebula located beneath the three stars that form Orion's Belt is bright enough to see with the naked eye. The Orion Nebula, also known as M42, is a gas cloud lit up by hot stars embedded within it. Radiation from these stars "excites" (energizes) gases in the nebula*

nebulae are billowing clouds of gas and dust

KEY WORDS

INTERSTELLAR MATTER: matter found in the space between the stars

NEBULA: a cloud of gas and dust among the stars

and makes them glow. This kind of nebula is called an EMISSION NEBULA. Most emission nebulae give out red light, which is the color given out when hydrogen gas (the most common gas in nebulae) becomes excited.

"What do you think is between the stars?"

* There are also REFLECTION NEBULAE, which give out a whiter light as they simply reflect the light from nearby stars, and DARK NEBULAE, which are not lit up at all. We see the latter just as dark patches against a starry background, when they blot out the light from stars behind them. There are many of them in the Milky Way, where they look like holes in the night sky. One, in the Southern Cross, is appropriately called the Coal Sack.

* Astronomers have identified more than 50 different substances in nebulae, among them glycine, an amino acid that is one of the building blocks of life. ***The inference is that life could be comparatively common in the universe.***

MESSIER'S NUMBERS

The Orion Nebula is known as M42 because it was number 42 in a catalog of nebulae compiled in the 18th century by the French astronomer **Charles Messier** (1730–1813). Like other astronomers of his time, Messier got tired of mistaking nebulae for comets. **He therefore proceeded to draw up a list giving the location of nebulae, so their confusion with comets could be eliminated.** By 1784, he had compiled a list of 103 objects, six more being added later. As well as true nebulae, he included galaxies and star clusters, which looked like misty patches in the telescopes of his day. **It was not until early this century that astronomers were able to distinguish between true nebulae and galaxies, which they at first called extragalactic nebulae, meaning nebulae outside our own galaxy.**

A STAR IS BORN

* Stars are born, grow up, and die of old age, just like living things. But their life spans are measured in billions or tens of billions of years. Stars are born in nebulae, the great clouds of gas and dust that exist in space. They start to shine when nuclear furnaces light up inside them.

KEY WORDS

BROWN DWARF:
a failed star, too small to shine

GIANT MOLECULAR CLOUD:
a region in which interstellar matter is denser than usual

STELLAR NUCLEAR FUSION:
the joining together of the nuclei (centers) of atoms of hydrogen due to the high pressures and temperatures in a star's interior, releasing abundant energy

BIRTH PANGS

* Star formation occurs only in certain regions of the universe, known as GIANT MOLECULAR CLOUDS, where interstellar gas and dust are denser than usual. No one really knows what triggers STAR BIRTH, but it could be the immensely powerful shock waves given out by a nearby exploding star,

the structure of an atom

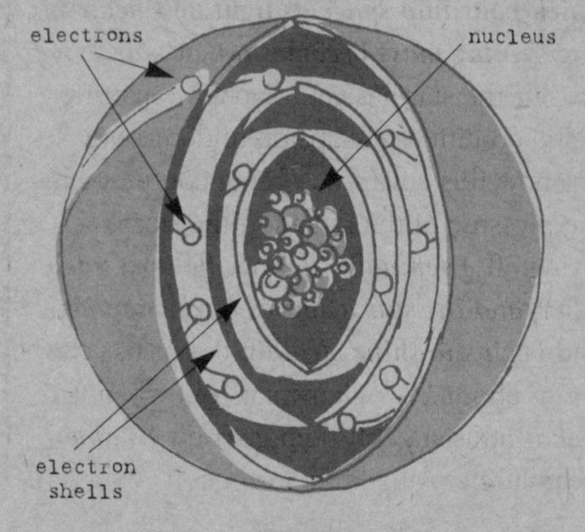

or supernova. ***Anyway, part of the nebula thickens up and starts to collapse under the mutual gravitational attraction of its particles. As the cloud collapses, forming a denser and denser mass, it releases energy and its temperature rises. At the same time, the mass begins to swirl like a whirlpool.***

LIGHT THE FURNACE

* As time goes by—usually tens of millions of years—the mass gets smaller and smaller, denser and denser, and hotter and hotter, spinning all the while. This so-called PROTOSTAR begins to glow dully, from time to time blowing off matter into space. Eventually, the temperature in its core reaches 10 million degrees or more. ***At this point a sudden and dramatic change occurs.*** The hydrogen particles that make up the bulk of the protostar begin to fuse together to make a new element, helium. ***This process of nuclear fusion in the core liberates fantastic amounts of energy, which pour into space as light and heat. The shining mass becomes a new star.***

* But the star has yet to achieve maturity. It still continues collapsing under gravity, but now this collapse is being countered by the pressure of the radiation the star is giving off. ***Eventually, the two balance each other, and the star achieves a constant size and begins to shine steadily.*** A star like the sun is about 50 million years old when it reaches this state. It then goes on shining steadily, changing little, for billions of years.

The main sequence

A mature and steadily shining star is called a main-sequence star, as it occupies the broad main-sequence band of the Hertzsprung-Russell diagram (page 77).

it's that collapsing matter again

BROWN DWARFS

Not all protostars achieve stardom. If the mass of the collapsing nebula is too low, the temperature inside the protostar never rises high enough to trigger nuclear reactions. Instead, it becomes a **brown dwarf**, glowing feebly before gradually fading away.

I'm getting a bit old

DYING SUNS

stars like the sun live 10 billion years

* When stars reach the end of their lives, some exit the stellar stage with a bang, others with a whimper. It all depends on their mass. A relatively low-mass star like the sun bids farewell with a whimper—ending up as a tiny body little bigger than the Earth.

SWAN SONG

* ***Born about 4.6 billion years ago, the sun is at present middle-aged; stars like it live for about 10 billion years***. Then they start to die. This happens when the hydrogen fuel in their core runs out. All that is left is the product of fusion, helium. ***With no outgoing radiation from fusion to battle against, gravity takes over and causes the core to start to collapse***. The collapse releases energy that heats up the outer gas layers and makes them expand. ***The star begins to swell until, typically, it is tens of times larger than its original size, becoming what astronomers call a*** RED GIANT, ***from the fact that its cooler outer surface gives off red light.***

when all the hydrogen runs out we are left with helium

KEY WORDS

BLACK DWARF: the tiny blackened corpse of a star like the sun

PLANETARY NEBULA: a ringlike cloud of gas around a dying star, which through a small telescope looks a little like a planet

RED GIANT: a dying star that has expanded greatly from its original size and gives off red light

WHITE DWARF: a tiny, dense, hot star, representing a late stage in the life of a star like the sun

GRAVITY THE WINNER

★ The star's core goes on shrinking and heating up. ***At a temperature of about 100 million degrees, more fusion reactions are triggered (this time between helium nuclei).*** These produce carbon, and release energy that keeps the red giant shining. ***But in time the helium runs out, too. Then gravity takes over and the giant begins to collapse.***

Heavyweight

*The matter in a **white dwarf** is so incredibly dense that a single teaspoonful of it would weigh several tons.*

the life of a sun-like star

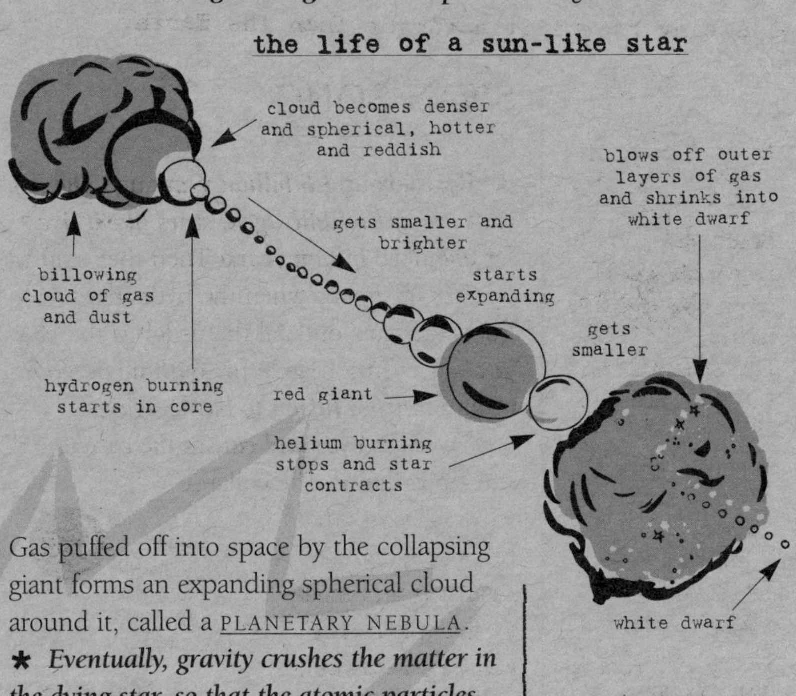

Gas puffed off into space by the collapsing giant forms an expanding spherical cloud around it, called a PLANETARY NEBULA.

★ ***Eventually, gravity crushes the matter in the dying star, so that the atomic particles are packed tightly together. The star is now no longer a giant, but a*** WHITE DWARF ***a few thousand miles across.*** At this stage it is still hot and shining brightly. But as time goes by, it cools and dims until it becomes a BLACK DWARF ***and, ultimately, disappears into the inky blackness of space.***

"Oh no, it's a catastrophe!"

STELLAR CATASTROPHE

★ Stars that are several times more massive than the sun have a much shorter life span (millions rather than billions of years). They, literally, end their lives with a bang—the biggest bang in the universe—and for an instant may shine with the brightness of a billion suns.

a supergiant exits early—with a bang, not a whimper

The Crab Nebula

In AD 1054 Chinese astronomers recorded the appearance of a "new star" in the constellation Taurus. It was, in fact, a supernova. In its place today a still-expanding cloud is visible, known as the Crab Nebula because of its shape.

SUPERGIANT

★ *Like their smaller relatives, massive stars expand into red giants (see page 92) when they use up the hydrogen fuel in their core and begin to die. But then they keep on expanding until they are hundreds of times their original size, becoming what are called* SUPERGIANTS. Like ordinary red giants, supergiants have a relatively cool surface and give out reddish light.

ELEMENT FACTORY

★ *The energy that's required to keep a supergiant shining comes from a whole series of nuclear reactions taking place in the star's core*. Instead of nuclear fusion stopping with the helium–carbon reaction, as happens in a red giant, it continues with carbon fusing to form heavier elements, such as magnesium. In other parts of the core, fusion reactions produce a host of other elements, such as silicon, sodium, sulfur, nickel, and iron.

supergiants feed voraciously...

SUPERNOVA

✱ *The supergiant feeds voraciously on its nuclear fuel, and within a few thousand years (a mere blink of a cosmic eye) there is none left. First the core and then the rest of the enormous stellar mass collapse under gravity. The collapse is so rapid and releases so much energy that a gigantic explosion takes place and the star is blown to bits*. During this explosion—called a SUPERNOVA—the star blazes like a beacon and may for an instant be brighter than all the stars in its galaxy combined.

✱ *The matter blasted into space by the supernova forms a cloud, which expands as time goes by. Many of these* SUPERNOVA REMNANTS *are visible in the heavens. In recent years the Hubble Space Telescope has been observing the expanding remnant of the supernova that took place in 1987 in our neighboring galaxy, the* LARGE MAGELLANIC CLOUD *(see page 102).*

KEY WORDS

SUPERGIANT:
a huge star, hundreds of times bigger than the sun

SUPERNOVA:
the mammoth explosion that takes place when a supergiant dies

SUPERNOVA REMNANT:
an expanding cloud formed by matter ejected during a supernova explosion

a supernova explosion

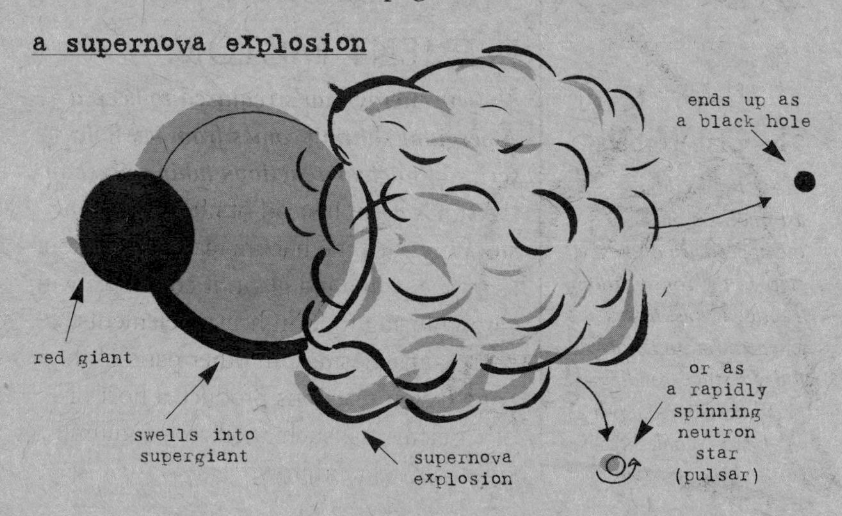

on off on off

CELESTIAL LIGHTHOUSES

* When a massive star explodes as a supernova and blasts itself apart, the core remains and starts to collapse rapidly under gravity. In some stars, the collapse halts when the core has shrunk to a dense city-sized object. This flashes pulses of energy into space, like a celestial lighthouse.

pulsars flash like lighthouses in space

Long time no see

The supernova in the Large Magellanic Cloud in 1987 was the brightest seen on Earth for nearly 400 years. Even though it occurred 70,000 light-years away, it was visible to the naked eye.

NEUTRON STARS

* Stars with a mass of about 1½ to 3 times that of the sun end up as tiny pulsating objects, or PULSARS, when they die. ***Astronomers think that pulsars are rapidly spinning*** NEUTRON STARS ***(stars made up only of neutrons). In them, gravity has squashed the neutrons tightly together, creating what is known as*** DEGENERATE MATTER. ***Typically, a neutron star measures only about 12 miles across, yet it has maybe twice the mass of the sun.*** Its density is unbelievably high—millions upon millions of times that of lead.

avoid degenerate weightlifting (neutron stars are millions of times denser than lead)

MAGNETIC MAYHEM

★ No one is really sure exactly how a spinning neutron star flashes pulses of energy into space. But it certainly has to do with the star's powerful magnetic field. ***Electrons moving rapidly in a magnetic field give off radiation, and it may well be that in a pulsar this radiation escapes as a beam through holes in the magnetic field over the magnetic poles***. As the star rotates, the beam sweeps across our line of sight as we view the pulsar from Earth, and we see a pulse—in the same way that we see a flash from a lighthouse when its beam sweeps past our eyes.

"Bet they don't know what we are!"

SPEEDY FLASHERS

★ ***The first pulsars found emitted energy as radio waves, but there are others that emit X-rays and light too.*** The one in the Crab Nebula resulting from the supernova explosion of AD 1054 (see page 94) spins around and flashes 30 times a second. But compared with some pulsars, that's slow. ***One of the*** MILLISECOND PULSARS, ***known as 1937+21, flashes 642 times a second.***

ET calling?

*Astronomer **Jocelyn Bell** discovered the first pulsar in 1967, while working at the Mullard Radio Astronomy Observatory, in Cambridge. But at the time, no one knew what it was. Every 1.3373 seconds the strange radio source sent out radio pulses about one twentieth of a second long. It seemed like a radio signal from some far-off alien world, so the astronomers dubbed it LGM–short for Little Green Men.*

KEY WORDS

DEGENERATE MATTER: matter in which atomic particles (that are separated in ordinary atoms) are squashed together

MAGNETIC FIELD: the region around a star (for example) within which its magnetism acts

PULSAR: a rapidly spinning neutron star that sends out beams of radiation

"I feel pulled in all directions!"

THE ULTIMATE ABYSS

★ When a really massive star collapses, the end result is not a neutron star, or anything that can actually be seen. All that remains is a region of space with incredibly high gravity. Astronomers call it a black hole, and it is the most fascinating and awesome thing there is in the whole of the universe.

Spaghettification

If you are unfortunate enough to fall into a black hole, the prognosis isn't favorable. Gravity within the black hole increases at an alarming rate. So it would exert a much greater pull on your feet than your head. You would end up very long and very thin, like spaghetti.

Name dropper

US physicist ***John Wheeler*** *first used the term "black hole" for a completely collapsed star at a meeting at the Institute for Space Studies, in New York, in December 1967.*

A BLACK HOLE IS BORN

★ *Some stars are truly massive—tens of times more massive than the sun. They lead brilliant but relatively short lives, lasting just tens of millions of years. When such a star dies, the collapse of its stupendous mass under gravity is inexorable.* No force in the universe is able to resist the pressure

keep away from black holes, they're terrible suckers

of the in-falling matter, not even the nuclear forces within the atom. The star becomes ever smaller, and its gravity ever greater.

NOW YOU SEE IT...

* ***Eventually the star shrinks to such a minute size that its gravity is too high for anything—not even light rays or other radiation—to escape from it.*** At this point the star's radius is known as its SCHWARZSCHILD RADIUS. For a star with a mass of 10 suns, it's about 18 miles.

"Where are all those black holes?"

fascinating, awesome and invisible...

* Because light now can't escape from the shrunken star, the star disappears from the observable universe. It becomes a BLACK HOLE in space. Anything that goes close to it will get sucked in and won't ever get out.
* The boundary between the invisible black hole and the visible universe is known as the EVENT HORIZON. ***What happens inside the event horizon can never be known. But astronomers reckon that the collapsing star continues to shrink until it becomes compressed into an infinitely small and infinitely dense point, known as*** a SINGULARITY. Such a concept stretches comprehension to the limit.

FINDING BLACK HOLES

Although we can't see black holes, we can sometimes find out where they are. **This is possible when a black hole is part of a two-star (binary) system in which the other star can be seen. In such a system, the prodigious gravity of the black hole tends to suck matter away from the visible star.** As this matter spirals towards the black hole, it forms an **accretion disk** near the event horizon. Friction within the disk heats up the swirling matter and causes it to emit energy as X rays, before it disappears into the hole. When astronomers spot X ray emissions from a system of this kind, they infer that a black hole is present. For example, the powerful source of X rays in the constellation Cygnus, known as Cygnus X-1, probably comes from a black hole companion to the supergiant star that is visible.

CHAPTER 4

THE GALAXIES

* Stars are not found scattered around in space haphazardly. In fact, they congregate into vast star islands. The space in between these islands is star-free.

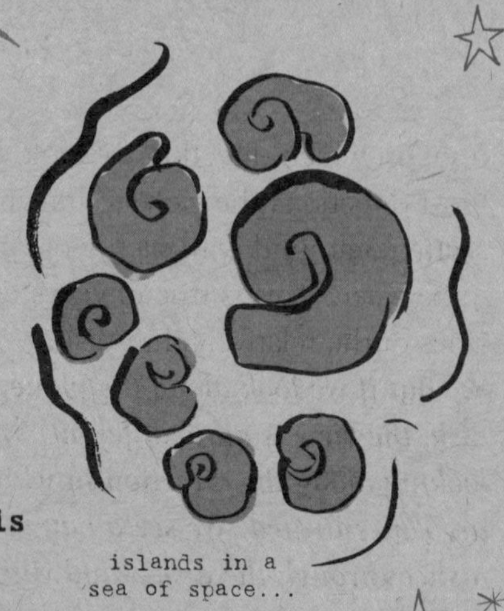

islands in a sea of space...

"It's good to be back in the home galaxy!"

KEY WORDS

GALAXY:
a star island in space
THE GALAXY:
our home galaxy (also known as the Milky Way Galaxy)

OUR GALAXY

* These star islands are what we call galaxies. ***The sun and all the other stars we see in the night sky belong to our home galaxy, which astronomers usually refer to as*** THE GALAXY (with a capital "G"). The nebulae and star clusters we see in the sky belong to the Galaxy, too.

* ***The Galaxy takes the form of a*** DISK ***with a bulge in the center—it has been described, inelegantly, as being like two fried eggs stuck back to back. In the flattened part of the disk, stars are found gathered into*** ARMS ***that spiral out of the thicker center (the nucleus).*** The sun is part of what is called the ORION ARM and lies about 30,000 light-years from the center of the Galaxy, which in total measures some 100,000 light-years across.

THE VIEW FROM INSIDE

* Because the sun—and Earth—are inside the disk, what we see in the heavens depends

upon in which direction we look. In most directions the view is similar—stars scattered around with plenty of dark space in between. This is true of views out of the sides of the relatively thin disk.

* *But if we look along the plane of the disk, the view is quite different. We are looking along the direction in which stars are concentrated. We see a band—in effect a slice through the disk—and within it stars beyond number seemingly packed close together.*

* *This is the band we see in the heavens as the Milky Way*. It resolves into closely packed stars when we look at it with a telescope. *As a result, the Galaxy is sometimes called* the MILKY WAY GALAXY.

"It depends on the way you look at it!"

look along the plane to view the Milky Way

* *The Galaxy contains some 100 billion stars, concentrated on the spiral arms. The whole Galaxy spins around slowly, so that from afar it would look like a rotating pinwheel.*

Center of the Galaxy

*One of the brightest regions of the **Milky Way** is in the constellation **Sagittarius**. The stars are packed very closely together here, forming dense star clouds. There is a good reason for this—**the galactic center** lies in this direction. We can't see right into the center, however, because of obscuring dust clouds in between.*

Sagittarius the Archer

The rotating pinwheel

*It takes the sun about 225 million years to circle around the galactic center, a period astronomers refer to as a **cosmic year**. The strange thing is that the dense balls of stars we call globular clusters don't take part in the general rotation of the Galaxy. They pursue independent orbits around the center, forming a kind of halo around the Galaxy.*

OUTER GALAXIES

★ Our Galaxy is just one of many billions of star islands that make up the universe. While many of the other galaxies are like our own, others are very different. Even though they are very far away, we can see a few of them with the naked eye.

Seeing the Whirlpool

The third Earl of Rosse (1800–67) was the first person to spot a spiral galaxy, though at the time it was thought to be a nebula (see page 42). He discovered it in 1845, using a huge 72-inch reflector that he had built at his home at Birr Castle, in Ireland. Appropriately called the Whirlpool, it was the galaxy M51.

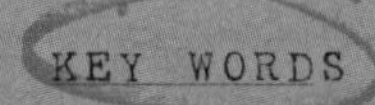

KEY WORDS

ELLIPTICAL:
a galaxy that has an oval or round shape

SPIRAL:
a galaxy that has its stars on curved arms

IRREGULAR:
a galaxy that has no particular shape

THE GREAT NEBULA

★ In the constellation Andromeda, not far from the easy-to-spot Square of Pegasus, is a faint misty patch. ***Until early this century it was called the Great Nebula in Andromeda, because it was thought to be part of our Galaxy. We now know that it is an*** EXTRAGALACTIC NEBULA ***(see page 89)***. Like other outer galaxies, it contains a mixture of stars, clusters, and nebulae similar to our own. Although it lies more than 2 million light-years away, the GREAT NEBULA or ANDROMEDA GALAXY is one of our closest galactic neighbors.

MAGELLAN'S CLOUDS

★ ***But the two galaxies known as the Large and Small Magellanic Clouds are right on our cosmic doorstep. In the far Southern Hemisphere, you can see these two misty patches (in Dorado and Tucana) with the naked eye.*** They are named after the famed Portuguese navigator, Ferdinand Magellan, who observed them. ***The*** LARGE MAGELLANIC CLOUD ***is the closer of the two, being a mere 170,000 light-years distant from the Earth.***

"So where is Andromeda?"

TYPES OF GALAXY

★ The Andromeda Galaxy is like a larger version of our own Galaxy—with arms spiraling out of its nucleus. We call this type a SPIRAL GALAXY. In contrast, the Magellanic Clouds have no special shape and are classed as IRREGULARS. The two other main types of galaxy are BARRED SPIRALS and ELLIPTICALS. Barred spirals have curved arms that extend out of the ends of a bar through the nucleus. Ellipticals are oval to spherical in shape and lack the arms of the spirals; they make up about three-quarters of all the galaxies in the universe.

the Square of Pegasus and the Andromeda Galaxy

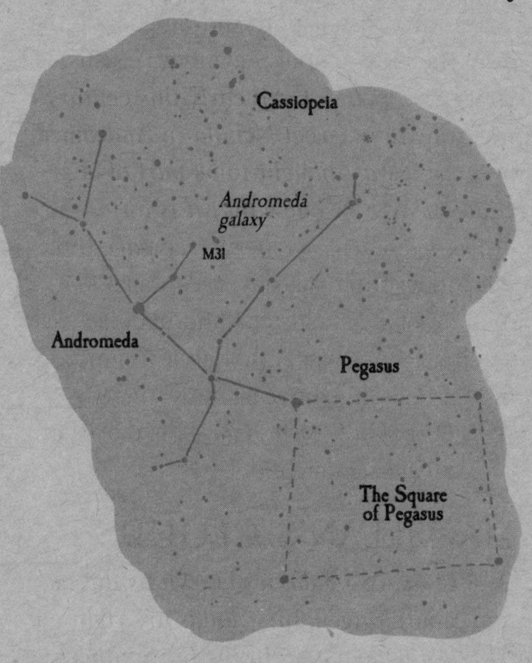

...not far from the Square of Pegasus

THE HUBBLE SYSTEM

We still classify galaxies using the method formulated in the early 1900s by Edwin Hubble (1889–1953), which splits the three main types into categories. Ellipticals are classed as E0 to E7, according to their shape. Spirals (S) and barred spirals (SB) are typed a, b or c, depending on how open the spiral arms are. Our own Galaxy has reasonably well-spaced arms and is therefore classed as an Sb—though there is some evidence of a bar through the nucleus, so it could be an SBb.

GETTING HYPERACTIVE

* Nine out of ten galaxies are "normal," in that we see them by the light of their billions of stars. A few galaxies, however, give out much more energy than usual—often not just as light.

ACTIVE GALAXIES

* *In the 1940s the American astronomer* CARL SEYFERT *(1911–60) began studying galaxies possessing a particularly bright nucleus.* Today, many of these SEYFERT GALAXIES are known. *They are just one type of* ACTIVE GALAXY*—a galaxy that has a prodigious energy output derived not solely from starlight. In Seyfert galaxies the energy seems to be produced by madly swirling clouds of hydrogen gas in the nucleus.*

* *In the 1950s radio astronomers began finding galaxies that were pumping out enormous energy as radio waves, rather than light.* One of the first found, Cygnus A, has about 10 million times the radio output of an ordinary galaxy. *The strange thing about* RADIO GALAXIES *is that the radio waves don't appear to come from the galaxy itself, but from regions of space that lie millions of light-years on either side.*

KEY WORDS

ACTIVE GALAXY:
one with an exceptional energy output

QUASAR:
a starlike object with the output of hundreds of galaxies

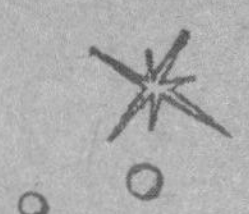

QUIRKY QUASARS

* *In the 1960s radio astronomers began matching visible objects with the powerful radio sources they were picking up. In 1962 they identified the source known as 3C-273 (number 273 in the third Cambridge catalogue of radio sources) with what looked like a star. By checking its spectral lines (see pages 74–5), they found it had a red shift that placed it an unbelievable 2 billion light-years away. At this distance no ordinary star would be visible. And for any object to be visible at this distance, it would have to have the combined energy output of hundreds of galaxies!*

* Since then, thousands of such bodies have been discovered. They are known as QUASARS (short for QUASISTELLAR RADIO SOURCES) if they give off powerful radio waves, or QSOs (QUASISTELLAR OBJECTS) if they are "radio quiet." They are all very remote. Indeed, some seem to be right at the edge of the observable universe, more than 13 billion light-years away. *Or to put it another way, when we see quasars we are looking back to a time when the universe was very young.*

"Is it the Beatles or could it possibly be a quasar?"

THE POWER HOUSE

Seyfert galaxies, radio galaxies, quasars, and other active galaxies called blazars must clearly have an exceptional power source, or "engine," to generate such fantastic amounts of energy. **Astronomers reckon that the only engine anywhere near powerful enough is a black hole (see page 98). Enormous amounts of energy are produced inside the rapidly spinning accretion disk (see page 115) that surrounds a black hole—and because the accretion disk is thick and surrounded by a doughnut-shaped torus of gas and dust, the energy is not able to pass through it. So it escapes, instead, as an intense jet of radiation along the spin axis.** The images of jets and torusi in active galaxies sent back by radio telescopes and the Hubble Space Telescope support the black-hole engine theory.

GALAXIES AND THE UNIVERSE

* Astronomers reckon that there are probably as many as 15 billion galaxies in the universe. But, like stars, they are not found scattered about haphazardly. They are found together in groups both on a small and a large scale.

the voids in the superclusters interconnect, giving the universe a spongy structure

KEY WORDS

SUPERCLUSTER:
a grouping of hundreds of clusters of galaxies

VOID:
a region of space between superclusters that contains no galaxies

A LOCAL GROUP

* The Large and Small Magellanic Clouds are more than neighbors of our own Galaxy, they are satellites that orbit around it. ***The Galaxy also forms part of a larger grouping of at least 20 galaxies, known as the*** LOCAL GROUP. The Andromeda Galaxy and another spiral, M33 (in the constellation Triangulum), also belong to the Group. It occupies a region of space about 3 million light-years across.

CLUSTERING TOGETHER

* The Local Group is just peanuts compared with the CLUSTER of thousands of galaxies in the constellation Virgo. ***The Virgo Cluster sprawls across about 7 million light-years of space, and has at its heart giant elliptical galaxies nearly as big as the Local Group itself. At a distance of 50 million light-years or so, it is the nearest big cluster.***

★ There is another large cluster in the constellation Coma Berenices, 350 million light-years away, again made up of thousands of galaxies with giant ellipticals at its heart. ***In fact, giant ellipticals are found at the center of many clusters, where they were probably formed by successive collisions and mergers of smaller galaxies.***

the Great Wall is the largest supercluster known

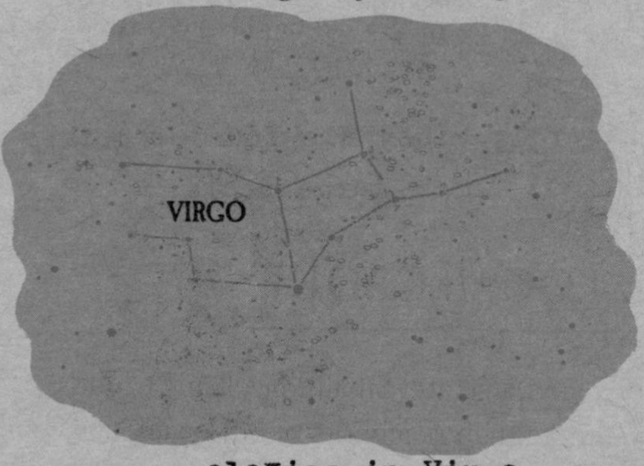

galaxies in Virgo

REALLY SUPER

★ Galaxy clusters also group together into SUPERCLUSTERS. ***The Virgo Cluster and the Local Group both form part of the Virgo (or Local) Supercluster, which extends into Leo, Crater, and Canes Venatici. It includes several hundred clusters, in a region of space about 100 million light-years across.***

★ As in other superclusters, the galaxies seem to be arranged in curved sheets around roughly spherical regions of space called VOIDS, where there are no galaxies at all.

★ ***The voids in the superclusters connect with one another, giving the universe a kind of spongy structure.***

THE GREAT WALL

The largest supercluster that's known to astronomers today takes the form of a "crumpled membrane" bearing countless numbers of galaxies. **Nicknamed the Great Wall, it is the largest structure discovered in the universe so far.** The area that it covers is approximately 280 million by 800 million light-years—but it is comparatively shallow, only about 23 million light-years thick.

THE EXPANDING UNIVERSE

★ With just a few local exceptions, all the galaxies we see in the heavens appear to be rushing headlong away from us. And the farther they are away, the faster they appear to be traveling. The implication is that the whole universe is expanding, as if from a mighty explosion eons ago.

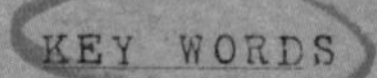

BIG BANG:
the event that is thought to have created the universe

COSMOLOGY:
the study of the origin and evolution of the universe

"Yes—it's expanding!"

the Big Bang must have been really violent

BIG BANG

★ *If the universe is expanding, it follows that it must have been smaller in the past. And there must have been a time when all the matter that is currently in the universe was packed together in one spot—theoretically in a spot that was infinitely small. By observing the rate at which the universe appears to be expanding today and working backward, astronomers have estimated that this must have been the case about 15 billion years ago.*

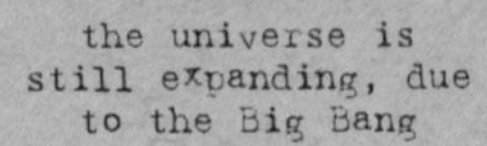

the universe is still expanding, due to the Big Bang

★ The universe, they say, came into being around that time, in an event visualized as an almighty explosion, known as THE BIG BANG. ***In the Big Bang theory, all things that exist—matter, energy, space, the forces of nature—were created. We can't ponder about what happened before the Big Bang, because time itself was created then.***

THE FIRST SECONDS

★ No one knows how or why the Big Bang happened—a divine creation, perhaps. But astronomers think they know what the universe was like just a few millionths of a second afterward. And they have worked out how the universe has evolved from then until the present day. The study of the origin of the universe and how it evolved is known as COSMOLOGY. Cosmologists and physicists have pooled their knowledge to envisage conditions in the newborn universe.

★ ***Moments after the Big Bang, the universe is incredibly hot and full of*** PHOTONS, ***or "packets" of radiation. But there is no matter. As the universe cools, radiation begins to turn into matter (Einstein's $E = mc^2$ again), creating the first subatomic particles with which we are familiar***—PROTONS, NEUTRONS, ***and*** ELECTRONS. ***At this stage, the universe is just 10 seconds old.***

The next minutes

Within minutes, as the new universe expands and cools further, the particles start combining together: protons merge with neutrons to form helium nuclei. Present now in the universe are photons of radiation, protons, helium nuclei, and electrons.

right place, wrong time

if we traveled back in time we could see what happened when the universe began

As time goes by

After 300,000 years the universe has cooled to 5,500–7,250°F, allowing particles to combine to form atoms. The fog of particles clears, leaving the universe transparent.

THE PRESENT AND FUTURE UNIVERSE

★ Investigation of our present universe has revealed convincing evidence that a Big Bang caused the universe to expand initially. But it is far from clear what will happen to the universe in the future.

"I see overall gravitational attraction and eventual collapse!"

FIREBALL RADIATION

★ *From the moment the universe became transparent, heat radiation has been coursing through it—assuming, of course, our theory about the Big Bang is correct. By now, astronomers estimate, it should give the whole universe a temperature of*

fireball radiation coursing through the universe

In the Crystal Ball

Will the universe continue expanding forever? Or will the expansion stop one day? Lacking any suitable crystal ball, cosmologists confess they don't know what will happen to the universe eventually. **The problem of working out what will happen to it hinges on how much mass it contains.** If there is enough mass, its overall gravitational attraction will one day halt the expansion of the universe and eventually cause it to collapse. All its matter might come together in a reversal of the Big Bang, in an almighty **Big Crunch**.

about three kelvin (3K)—in other words, three degrees above absolute zero. In 1965 two Bell Laboratories scientists, ARNO PENZIAS and ROBERT WILSON, picked up some background radiation in space that matched this temperature. ***It was a triumph for Big Bang cosmologists—and proved to be a cosmic nail in the coffin for those who supported a radically different theory about the universe known as*** THE STEADY STATE THEORY.

STEADY AS SHE GOES

* The vociferous proponents of the Steady State theory were FRED HOYLE, HERMANN BONDI, and THOMAS GOLD, who advanced the theory in 1948. ***The universe was essentially the same in the past, they said, as it is today. Galaxies form and recede, but to keep the universe in a steady state, new matter is continually being created in space. In time this forms into galaxies, which in their turn recede, and so on. If we are prepared to accept the idea of matter being created (in one fell swoop in a Big Bang), they argued, why couldn't it be created little by little?***

Will the universe end in a Big Crunch?

Is the Universe Open or Closed?

The idea of a Big Crunch envisages a ***"closed"*** *universe with definite boundaries. If there isn't enough mass to make the universe collapse onto itself, then the universe will continue to expand forever–it will be an* ***"open"*** *universe. Adding up the mass of all the galaxies thought to exist, there doesn't seem to be sufficient mass for a closed universe.* ***But there is plenty of dark matter in the universe–matter we can't see.*** *For example, it is now known that particles called* ***neutrinos,*** *once thought to be massless, do in fact have slight mass.* ***Since they are produced in prodigious quantities in the nuclear furnaces of the stars, neutrinos could conceivably tip the balance toward a closed universe and a Big Crunch.***

CHAPTER 5

THE SUN'S FAMILY

★ The Solar System is like a great celestial merry-go-round—with a diverse collection of heavenly bodies whirling around the sun. There are thousands of such bodies, but they are so widely scattered that the solar system consists mainly of empty space.

a 17,400 mph space shuttle would take at least 50 years to cross the solar system

THE ORBITS

★ The planets all circle around the sun in the same direction. *Viewed from cosmic north, they travel counterclockwise. This is the preferred direction of rotation in the solar system. The sun and most of the planets spin around on their axes in this direction, as do most of their moons.*

MEMBERS OF THE FAMILY

We shall be taking a look at the heavenly bodies that form the solar system in more detail later, but here is a brief introduction to the family of the sun.

SUN The star at the center of the solar system, and the only body within it that emits light of its own. Its enormous gravity holds the whole family together.

PLANETS The largest of the circling bodies that shine in our skies by reflecting sunlight. There are nine in all. Most have one or more moons circling around them.

MOONS Natural satellites of the planets. Only Mercury and Venus do not have moons.

ASTEROIDS Also called minor planets. Rocky bodies of all shapes, up to about 600 miles across.

COMETS Small lumps of ice and dust that shine when they approach the sun.

METEOROIDS Mainly small particles that sweep through interplanetary space. We see them as meteors when they plunge through Earth's atmosphere.

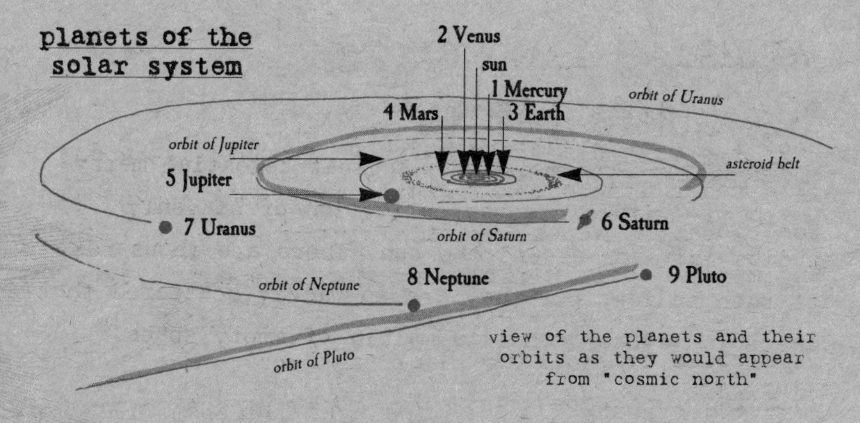

view of the planets and their orbits as they would appear from "cosmic north"

ECCENTRIC PLANETS

★ *Strictly speaking, the planets do not "circle" the sun—their orbits are oval, or elliptical, rather than truly circular.* But most of them don't deviate greatly from the circular. Only Mercury and Pluto show marked deviation, or ECCENTRICITY.

★ *Moreover, all the planets circle the sun in nearly the same plane*. Consequently, they appear to travel through the heavens in a narrow band, called THE ZODIAC (see page 67). ***Again, Pluto is an exception. The plane of its orbit deviates by more than 17 degrees from the norm.***

interplanetary travelers need to think long haul

SCALING THE SOLAR SYSTEM

The diagram (above) shows what the solar system would look like if you looked down at it from a point in the heavens near the Pole Star—from what we might call **cosmic north**. Notice how close together the four inner planets are compared with the outer ones, and that the "belt" of asteroids is located in between. **The scale of this diagram is almost impossible to comprehend. It spans a distance of some 7.5 billion miles. In a space shuttle traveling at the usual orbital speed (17,400 mph), it would take more than 50 years to travel from one side to the other.**

BIRTH OF THE SOLAR SYSTEM

* The solar system was not always the same as it is today. Born out of swirling clouds of gas and dust, it has taken more than 4½ billion years to reach its present state.

once its nuclear furnace ignited, the sun began to shine

Immanuel Kant had new ideas about the origin of the Solar System

NEBULAR IDEAS

The account of the formation of the solar system given here is often called **the nebular hypothesis**. It had its origins in the 1700s, in the ideas of German philosopher **Immanuel Kant** (1724–1804). These were developed further by the French astronomer **Pierre Laplace** (1749–1827).

BORN OUT OF PROPLYDS

* *Being a typical star, the sun condensed out of a great nebula. It started life as a shrinking ball of gas and dust that gravity had pulled together. As it shrunk, the gravitational energy that was released pushed up its temperature until it was hot enough for its nuclear furnace to light up. Then it began to shine.*

* By this time it was surrounded by a rotating disk made up of clumps of matter, called PLANETESIMALS. Near the inside of the disk, where the temperature was hottest, there were relatively heavy lumps of rock and metals, particularly iron. Farther out, where it was colder, there were lumps of ice and freezing gases. Farthest out, at the periphery, were concentrations of the lightest gases, hydrogen and helium. Astronomers call such a disk a PROPLYD,

standing for PROTOPLANETARY DISK. The Hubble Space Telescope has revealed many disks of this kind in star-forming nebulae.

GETTING TOGETHER

* As time went by, the small planetesimals whizzing around in the disk were constantly bumping into each other and merging together. ***Within a few million years, this process (called ACCRETION) had led to the formation of huge bodies that were, by their increasingly powerful gravitational attraction, swallowing up any nearby bits—like cosmic vacuum cleaners.***

INNER PLANETS

* *Four bodies—the inner planets Mercury, Venus, Earth, and Mars—formed relatively close to the sun. They were made up mainly of rock and iron. The energy released by the collisions that formed them caused them to heat up and melt. The heavy metals sunk to their center, forming the core of the planets as we know them today.*

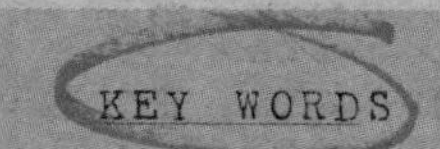

ACCRETION:
the formation of large bodies due to collisions with smaller ones

NEBULAR HYPOTHESIS:
the idea that the solar system was born out of a nebula

PLANETESIMALS:
the tiny lumps of matter from which the planets formed

Outer planets

*In the cold outer reaches of the newborn solar system, planets developed from the freezing gases. It is here that we now find the four **gas giants** Jupiter, Saturn, Uranus, and Neptune. These planets have a tiny core of rock, but they are mainly composed of gases. In thick atmospheres, the gases are found in their ordinary gaseous state, but in the interior of these planets they exist as liquids or even in solid form.*

SUN AND EARTH

* Justly regarded as a god by the ancients, the sun dominates—and is, in fact, responsible for—our lives. Without the light and warmth that it pours onto our planet, the Earth would be a colorless, deep-frozen and lifeless ball of rock.

the sun played a major role in Druid rituals

WARNING!

Never look directly at the sun. This will damage your eyes. And never look at the sun through binoculars, nor through a telescope. Doing so can blind you.

KEY WORDS

CLIMATE: the weather conditions generally prevalent in a region

TIME ZONE: a region in which there is a standard time

SUN IN THE SKY

* Every day the sun rises in the east, and arcs westward and upward till it reaches due south at noon. It then descends, traveling westward, until it disappears below the western horizon at sunset. *This diurnal (daily) motion of the sun defines the day, which is our primary unit of time.*

* *But the notion that the sun travels around the Earth is an illusion. It is, of course, the Earth that moves, not the sun; the Earth spins on its axis once a day, rotating from west to east.*

the moving sun is an illusion

TIME'S WINGED CHARIOT

★ In solar time, which we use for purposes other than astronomy (see page 62), NOON is a defining moment that would seem to be ideal for setting our clocks. *But every place on Earth has a different time when the sun is highest in the sky, depending on longitude—so if we set our clocks by local noon, they would all register a different time!*

★ That is why we have that scourge of long-distance travelers, different time zones. *There are 24* TIME ZONES, *in each of which clocks are set to a* STANDARD TIME. *Each differs from its neighbor by an hour*.

★ The INTERNATIONAL DATE LINE, which for the most part runs along the 180° line of longitude, marks the place on the Earth's surface where a new calendar day is deemed to begin.

SUN AROUND THE WORLD

Because the Earth is round, the sun does not follow the same path as it journeys across the sky, in every part of the world. Near the equator, it climbs steeply through the sky until it is almost directly overhead at noon. But as you travel north or south from the equator, the sun does not climb as high; and, by the time you reach far northern or far southern latitudes, it stays close to the horizon. **The path of the sun through the sky has two noticeable effects on places at a particular latitude. It determines the length of daytime, when the sun is over the horizon; and it holds the key to the climate.** Temperature is the major factor in climate, and this depends primarily on the intensity of the sun's rays, **which in turn is determined by the height of the sun's trajectory through the sky.**

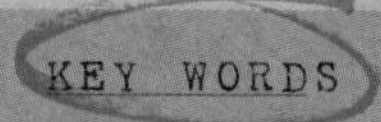

KEY WORDS

EQUINOX:
a time of the year when day and night are each 12 hours long

SOLSTICE:
a time of the year when the sun reaches its highest or lowest point in the sky

The solstices

Two significant turning points in the Earth's seasonal merry-go-round are the solstices, when the Earth's axis is tilted the most toward or away from the sun. In the Northern Hemisphere, the axis points the most toward the sun on about June 21 each year. This is the summer solstice, marking midsummer, although the weather hasn't fully warmed up yet. On about December 21, the axis points the most away from the sun, marking midwinter.

Stonehenge is aligned with the rising sun on Midsummer Day

THE SEASONAL SUN

★ The sun climbs higher or lower in the sky, according to the time of year. How high or low it is in the sky brings about the periodic changes in weather that mark the seasons.

EARTH'S TILTING AXIS

★ At the same time as the Earth spins on its axis, it is traveling in its yearly orbit around the sun. ***But the axis around which the Earth spins is not upright, or at right angles to the plane of its orbit. If it were, the sun would follow the same path across the sky all the time.*** And as we know, this doesn't happen. In summer the sun travels high; in winter, low.

* The reason this happens is because the Earth's axis is tilted at an angle of 23½ degrees with respect to its orbital plane. ***So, although the axis always points in the same direction in space, it points alternately toward and away from the sun at different positions in the Earth's orbit.***

midwinter—the Earth's axis is tilted away from the sun

* This means that a particular place on Earth is tilted more toward the sun at some times of the year than at others. ***When it is tilted more, the sun travels higher in the sky, and the weather is warmer. When it is tilted less, the sun travels lower, and the weather is cooler. The periodic warming and cooling as a result of the changing tilt reflect the time of the year, and define our seasons.***

... and in winter it is low

SEASONS DOWN UNDER

We have related the dates given for the seasons specifically to those in the Northern Hemisphere. In the Southern Hemisphere, the seasons are reversed —because when the Earth's axis is tilted most toward the sun in the Northern Hemisphere, it is tilted most away from the sun in the Southern, and vice versa.

The equinoxes

In between the solstices are the times of the year when the Earth's axis is tilted neither toward nor away from the sun. This happens around March 21 and September 23 each year. On both, day and night are the same length, which is why these dates are called the equinoxes, meaning "equal nights." March 21 marks the vernal (spring) equinox, when in the Northern Hemisphere the weather is starting to warm up. September 23 marks the autumnal equinox, when it is getting noticeably cooler.

I'm pretty indispensable for a yellow dwarf

OUR STAR, THE SUN

★ Boasting 750 times the mass of all the other bodies in the solar system put together, the sun dominates our corner of the universe. It is vitally important to us earthlings, but in the context of the universe as a whole it is an insignificant little star.

the sun is an averagely bright middle-aged yellow dwarf

KEY WORDS

FRAUNHOFER LINES: dark lines in the spectrum of sunlight

SUN DATA

Diameter:
865,000 miles
Volume (Earth = 1):
1,300,000
Mass (Earth = 1):
333,000
Density (water = 1): **1.4**
Average distance from Earth:
93,000,000 miles
Temperature of surface:
10,000 °F

THE YELLOW DWARF

★ *Because of the color of its light and its relatively small size for a star, astronomers call the sun a yellow dwarf* *(see page 77)**. It appears very much bigger and brighter than the other stars in the sky only because it is very much closer—millions of miles away, instead of millions of millions. Its yellow light signifies that it is an averagely hot star, with a surface temperature of about 10,000°F.*

sun gods always travel in style...

★ Like all stars, the sun is a great globe of hot gas that pours out fantastic quantities of light, heat, and other radiation. Only a fraction of the sun's energy output reaches us, but it is enough to give the Earth a pleasant environment for life to flourish. ***Like all stars, the sun produces its energy by means of nuclear-fusion reactions in its 23,400,000°F interior. Some 4 million tons of hydrogen are consumed every second in the mass-to-energy conversion process summed up by Einstein's famous equation $E = mc^2$ (see pages 44–5).***

WHAT'S IT MADE OF?

★ ***The sun consists of hydrogen (about 71 percent) and helium (27 percent), plus a cocktail of other elements (2 percent). In all, as many as 70 out of the 92 naturally occurring chemical elements have been found in the sun. Calcium, carbon, iron, magnesium, silver, and sodium are just a few of them.***

★ But how do we know this? From studying the spectrum of the sun's light—the same method astronomers use to find out the composition of other stars (see page 73). The sun's spectrum—the rainbowlike color band obtained by observing sunlight in a spectroscope—is crossed by many dark lines, called FRAUNHOFER LINES. ***The different sets of lines constitute the signatures of the various elements that make up the sun.***

FRAUNHOFER LINES

The dark lines in the sun's spectrum were first observed by the English astronomer **W. H. Wollaston** (1766–1828), in 1802, but are named after the German astronomer **Josef von Fraunhofer** (1787–1826), who first studied them in detail. Fraunhofer's work led to the development of that most useful astronomical instrument, the spectroscope.

the sun is a cocktail of as many as 70 chemical elements

oh no!

sun spots

THE STORMY SUN

* Dark spots, fiery fountains, gigantic flares, and rippling shock waves are just some of the awesome features we witness on the sun. They give us an insight into how other stars must behave.

the sun has a stormy temperament

KEY WORDS

AURORA:
colored lights seen in far northern and far southern skies
CHROMOSPHERE:
the sun's inner atmosphere
CORONA:
the sun's outer atmosphere
PHOTOSPHERE:
the sun's light-producing surface
SUNSPOT:
a dark, cool blemish on the sun's surface

The sun's crown

During a total eclipse, we can see the tenuous outer atmosphere of the sun, called the corona ("crown"), which is usually hidden by the photosphere's glare. It shows up as a pearly-white halo extending millions of miles into space.

THE SPOTTED SURFACE

* Light and other radiation pour into space from the sun's outer surface. ***Known as the* PHOTOSPHERE *("light sphere"), this is a layer about 250 miles thick that heaves like a stormy sea as pockets of hot gas keep bubbling up from deeper down***. This makes the surface look speckled, an effect known as GRANULATION. Periodically, the darker patches we call SUNSPOTS appear on the photosphere, usually in groups. They can grow to be thousands of miles across and may persist for months. ***About 2,700°F cooler than their surroundings, sunspots***

appear to be caused by strong magnetic fields. They come and go in cycles, peaking every 11 years or so.

THE COLOR SPHERE

* The sun's gaseous inner atmosphere is called the CHROMOSPHERE ("color sphere"), because of its reddish hue. Some 6,200 miles thick, it is in constant turmoil. *From it come great fountains of incandescent gas, known as* PROMINENCES, *hundreds of thousands of miles high. Leaping up in arching trajectories, they follow the magnetic lines of force surrounding the sun, which has a very intense magnetic field*. We aren't usually able to see these fiery fountains from Earth because of the glare of the photosphere. The only time we can do so is when the photosphere is blocked out by a TOTAL SOLAR ECLIPSE.

the northern lights are caused by the solar wind

A WINDY SUN

As well as radiation, the sun gives off a constant stream of electrified particles, such as electrons and protons. They flow out into interplanetary space, forming the so-called solar wind. At times of intense stormy activity on the sun, when explosive events called flares occur, the solar wind blows at gale force. When it reaches the Earth, it disrupts our atmosphere, causing electrical and magnetic storms that can knock out electricity supplies and cause disruption to radio communications. **The good news for the astronomer is that it creates the most beautiful shimmering curtains of color in the polar skies—the phenomenon known as an aurora.** In the Northern Hemisphere, they are called the **aurora borealis** or northern lights; in the Southern Hemisphere, the **aurora australis** or southern lights.

CHAPTER 6

THE PLANETS

★ Mercury is named after the fleet-footed messenger of the gods in Roman mythology—and aptly so. Of all the planets, Mercury takes the least time to travel around the sun, since it is the one closest to the center of the solar system.

Mercury, messenger of the gods

MERCURY DATA
Diameter at equator: **3,030 miles**
Average distance from sun: **36,000,000 miles**
Density (water=1): **5.4**
Spins on axis in: **58 days 16 hours**
Circles sun in: **88 days**
Moons: **0**

BRILLIANT MERCURY

★ ***Although Mercury is quite a small planet, it is both close enough to Earth and bright enough to be seen with the naked eye.*** At its most brilliant, it outshines Sirius, the brightest star in the sky. ***However, it is not easy to see because it always stays close to the sun and is therefore never seen in complete darkness. We see it either as a*** MORNING STAR, ***low on the horizon in the east just before the sun rises; or as an*** EVENING STAR ***in the west, immediately after sunset and again low down.*** Whether Mercury appears in the morning or in the evening depends on which side of the sun it happens to be, viewed from the Earth.

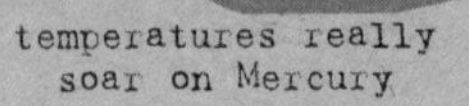

temperatures really soar on Mercury

THE LONG DAY

★ Like all planets, Mercury spins on its axis. But it spins very slowly. ***While the Earth spins around once a day, Mercury takes nearly two Earth***

months to spin around once. During this time it travels two thirds of the way around the sun. These relative motions create a situation in which Mercury's "day"—from one sunrise to the next—lasts for 176 Earth days. For the same reason, the planet's "night" is also 176 Earth days long.

one day on Mercury is the equivalent of 176 Earth days

TERRIFIC TEMPERATURES

★ ***During the long day on Mercury, the nearby sun beats down, baking the planet to temperatures of 800°F or more in places***. This is so hot that metals like tin and lead would melt. ***Conversely, of course, during the long night the surface soon cools down, and the temperature eventually falls as low as –350°F***.

THE CRATERED SURFACE

★ Through telescopes, we can make out only vague details on Mercury's surface. So astronomers had to wait until the planet was visited by the space probe *Mariner 10*, in 1974, before they learned what it looked like. ***In fact, it closely resembles the moon and is covered with craters, large and small. But it differs from the moon in not having any "seas," or "maria," as they are known. It has virtually no flat-plain regions. Mercury's most prominent feature is the huge circular Caloris Basin, probably caused by the impact of a huge meteorite in the distant past.***

INSIDE MERCURY

Mercury is classed as one of the terrestrial, or Earthlike, planets. Its density is similar to the Earth's, and it has a similar layered structure. But Mercury's iron core is relatively much larger, with a comparatively thin rocky mantle and crust on top. And its very thin atmosphere is hardly detectable.

KEY WORDS

EVENING STAR:
a planet that can be seen in the west immediately after sunset

MORNING STAR:
a planet that can be seen in the east immediately before sunrise

HELLISH VENUS

* Venus is a near twin of Earth in size, but the two planets have little else in common. It is named after the Roman goddess of love, but there is little that's lovable about the planet. Venus is a hellish place, with temperatures hotter than an oven and an atmosphere that can crush and suffocate.

temperatures on Venus are crushingly hot

VENUS DATA

Diameter at equator: **7,500 miles**
Average distance from sun: **67,000,000 miles**
Density (water = 1): **5.2**
Spins on axis in: **243 days**
Circles sun in: **224.7 days**
moons: **0**

"It's Venus, the goddess of love"

VENUS IN THE SKY

* *Apart from the moon, Venus is the brightest object in the night sky.* It outshines the other bright planets by a considerable margin: it has a maximum magnitude of –4.4, as opposed to –2.5 for Jupiter and Mars. *Like Mercury, Venus can be seen as an evening and a morning star. But its orbit takes it much farther from the sun than Mercury, which means we can see it much more easily. So when you see*

a bright "star" hanging in the western sky just after sunset, it will be Venus. And by the same token, if you are an early riser, that bright "star" shining just before sunrise will be Venus, too.

Venus looks so bright because it comes close to us (within 26 million miles) and has a covering of reflective cloud

SHOWING PHASES

* *As Venus orbits the sun, it appears to travel back and forth from one side of the sun to the other when viewed from the Earth. And we see different amounts of it lit up by the sun. In other words, it has phases, just like the moon does, and we see it as crescent, half-circle, or full disk at different points in its orbit. But, unlike the moon, the size of the planet seems to change with each phase, because it circles the sun inside the Earth's orbit*. It appears brightest at the half-circle phase, when it is at its maximum distance from the sun as viewed from the Earth. Astronomers call this its greatest ELONGATION.

THE HOTTEST BODY

If Venus ever had oceans like the Earth's, they would have evaporated long ago when the planet heated up. The only water present today is in the atmosphere. But the main gas in the thick, oppressive atmosphere (the pressure is 90 times greater than on Earth) is carbon dioxide. This heavy gas has been largely responsible for global warming on Earth, and on Venus it has created a galloping greenhouse effect. It traps most of the heat received from the sun and acts as a smothering blanket, stopping the heat from escaping back into space. **This greenhouse effect has made Venus the hottest planet in the solar system. Temperatures on its surface can soar to 900°F**. Adding to this hellish image, the clouds consist mainly of particles of concentrated sulfuric acid, which has a propensity for charring organic matter (including human flesh).

VENUSIAN LANDSCAPES

★ Until the Space Age was well under way, no one knew what our planetary neighbor and near-twin was like under its cover of cloud. Some suggested that the clouds on Venus concealed luxuriant vegetation, and maybe exotic wildlife of the dinosaur variety. The reality is very different. No vegetation, no life could tolerate the heat, and what we find is a scalding landscape of rolling rocky plains, towering volcanoes, and plenty of craters.

MAKING AN IMPACT

There are plenty of craters on Venus. Most mark the mouths of ancient volcanoes. Relatively few craters have been caused by the impact of meteorites. The main reason for this is that lava has flowed repeatedly over the surface and obliterated most of them. The ones that are visible are comparatively recent—probably no more than a few hundred million years old. This is very much younger than the majority of craters found on the moon.

THE INVASION OF VENUS

★ *The first successful US planetary probe, Mariner 2, gave us a hint of the extreme conditions on Venus*. Since that first hint, a succession of Russian and US probes have flown past Venus, landed on it, and gone into orbit around it. Detailed mapping of the planet began in 1978, when Pioneer-Venus 1 used radar to "see" through the dense clouds.

Pioneer-Venus 1 used radar to peer through the Venusian clouds

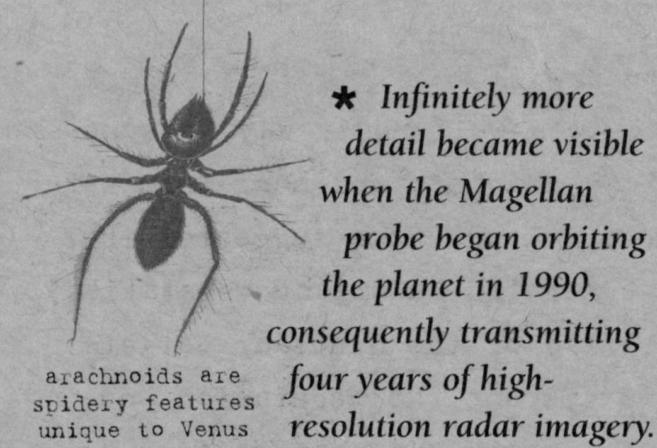

arachnoids are spidery features unique to Venus

★ *Infinitely more detail became visible when the Magellan probe began orbiting the planet in 1990, consequently transmitting four years of high-resolution radar imagery.*

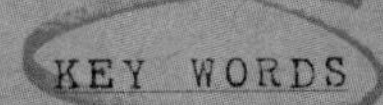

KEY WORDS

ARACHNOID:
a spiderlike feature on the surface of Venus

CORONA:
a crownlike feature on the surface of Venus

TWO CONTINENTS

★ *Venus consists mostly of gently rolling, low-lying plains. There are two main highland areas, or continents. The largest, Aphrodite Terra, runs along the Venusian equator, and is about the size of Australia. The other, Ishtar Terra, in the northern half, is about the size of South America.*

★ Like most features on Venus, the two continents have feminine names. They are, in fact, alternative names for the goddess of love. The ancient Greeks called her Aphrodite, the Babylonians Ishtar.

VOLCANOES AT WORK

★ Volcanoes are found all over the planet, and have spewed out lava to form great plains. Some volcanoes are quite small and cone-shaped. Others, thanks to repeated eruptions, have formed into great shield mountains, like those found on Earth. *One volcanic feature found on Venus definitely does not occur on Earth—circular lava "pancakes," typically about 15 miles across, which probably formed when lava bubbled up through surface cracks.*

Spiders and crowns on Venus

There are two other surface features that have no counterpart elsewhere in the solar system. *One is a spiderlike feature, called an* ***arachnoid****. It has a circular or oval center with cracks reminiscent of spider legs radiating out of it. Arachnoids were probably created by molten rock welling up from below and cracking the surface. The arachnoids may just be an earlier stage in the formation of the* ***coronae****, the planet's other unusual feature. These are crownlike structures, formed by closely spaced circular fractures, and no doubt result from interaction of welling-up lava with the surface crust.*

water ahoy!

PLANET EARTH

★ An alien astronomer looking at our solar system would single out the Earth as a particularly interesting planet—beautiful, wet, ever-changing, and alive.

EARTH DATA

Diameter at equator: **8,000 miles**
Average distance from sun: **93,000,000 miles**
Density (water = 1): **5.5**
Spins on axis in: **23 hrs 56 mins**
Circles sun in: **365¼ days**
moons: **1**

The water cycle

The sun evaporates water from the oceans into vapor that passes into the air. The vapor then cools and forms clouds of tiny water droplets. The droplets coalesce and fall from the clouds as rain, or snow if temperatures are freezing. Water has returned to the surface, ready to be evaporated back into the air.

THIRD ROCK FROM THE SUN

★ ***Situated between Venus and Mars, the Earth circles the sun at an average distance of about 93 million miles,*** a journey that takes a year. Accompanying the Earth as it speeds through space is its satellite, the moon. Like all the other planets, the Earth rotates on its axis as it travels through space. It spins around once in a day and 365¼ times in a year. ***But it does not spin in an upright position with respect to its path around the sun. Its axis is tilted at an angle of 23½ degrees, which brings about the seasons and their changing weather*** *(see pages 118–9).*

LEAP YEARS

★ The reason we have LEAP YEARS of 366 days, with a February 29 every four years (with occasional exceptions), is to allow for the ¼ day per year difference between the standard calendar year of 365 days and the natural year. ***Leap years are years that can***

be divided by four. The exceptions are century years that are not divisible by 400, which is why the millennium year 2000 is a leap year, but 2100 won't be.

PLANET WATER

★ Alien astronomers surveying our planet might prefer to call it "Water" instead of "Earth" because they would perceive that most of its surface is covered by watery oceans. *Oceans cover over 70 percent of the surface—the Pacific Ocean alone has a greater area than all the land regions, or continents, put together.*

the Earth is the only planet able to sustain life

★ The presence of water plays a vital role in the metabolism of the planet. *A never-ending exchange of water between the surface and the air dictates our weather.* This is known as the WATER CYCLE.

THE VITAL ATMOSPHERE

The atmosphere, or layer of air around the Earth, plays a vital role in permitting the one thing that makes the Earth unique among the planets—the existence of life. The atmosphere is made up of a mixture of gases, mainly nitrogen (78 percent) and oxygen (21 percent). It is the latter that almost all living things need to breathe to live. **The atmosphere also acts as a blanket in keeping the world warm at night, preventing too much of the heat of the day from escaping into space. In addition, it filters out from the sunlight invisible rays** harmful to living things. These include gamma rays, X-rays, and ultraviolet rays. Some of the latter get through—and tan us—but a layer of ozone in the atmosphere filters out most of them. However, atmospheric pollution is thinning the layer, creating so-called **ozone holes,** which are a cause for concern.

TO THE CENTER OF THE EARTH

★ Knowledge of the way the Earth is made up gives us an insight into the structure of other planets and their moons. It is made up mostly of rock, but has a heart of iron.

GEOLOGISTS AT WORK

★ *The Earth is the largest of the rocky planets. It measures 7,899 miles across at the Poles, but some 26 miles more at the equator, where its rotation makes it bulge.* Although on the outside the Earth is made up of solid rock (or rather different kinds of rock, see page 136), it isn't composed of the same rock all the way through. *Geologists tell us that the Earth is in fact made up of several layers, the outer one being the hard outer crust with which we are familiar.*

★ Studying the way that earthquake shock waves travel through the underground rocks reveals that the waves change direction at certain depths. This indicates that they are entering layers with differing compositions. The study of earthquakes is called SEISMOLOGY.

KEY WORDS

CORE:
the central part of the Earth

CRUST:
the hard outer layer of the Earth

GEOLOGIST:
a scientist who studies the Earth

MANTLE:
a thick rock layer between the crust and the core of the Earth

seismologists can have earthshaking careers

basic structure of the Earth

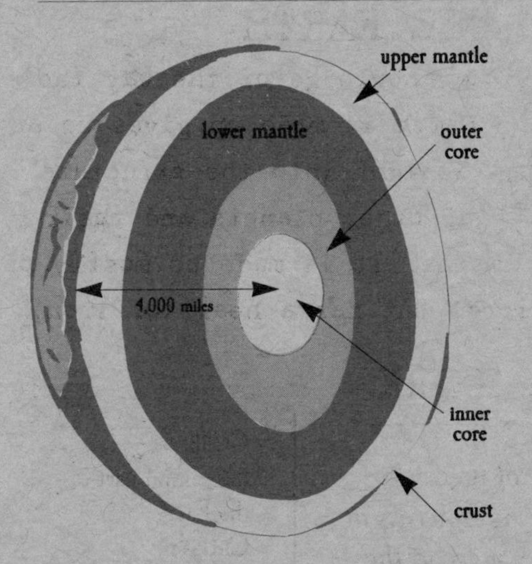

EARTH'S STRUCTURE

★ ***The CRUST is the thinnest layer of our planet. On the continents the crust is on average about 25 miles thick, but it is only about 6 miles thick under the oceans.*** Underneath the crust, and reaching down several thousand miles, is the MANTLE. In the upper part of the mantle is a layer called the ASTHENOSPHERE in which the rocks are very hot and nearly molten. They flow slowly, carrying the hard rocks on top with them, which explains the phenomenon of continental drift (see page 134).

★ Beneath the mantle is a two-layered CORE, made up mainly—perhaps surprisingly—of iron, with some nickel. These heavy elements must have sunk and settled in the center when the newborn Earth was still molten. ***The outer part of the core is still molten, even today.***

THE EARTH MAGNET

Because the Earth is spinning, currents ebb and flow in the molten outer core. Somehow, this turns it into a gigantic dynamo and produces electric currents. As always when an electric current flows in a conductor (which iron is) a **magnetic field** is set up. And this happens in the Earth. **The Earth is surrounded by a magnetic field that extends far out into space. It is this field that, among other things, causes compass needles to point north-south.** In effect, the Earth behaves as if it has a huge magnet embedded within it, with one of the magnet's poles situated near the geographic North Pole and the other near the geographic South Pole.

DRIFTING CONTINENTS

* The face of the Earth, as we know it today, will not stay the same forever. The continents are shifting and the oceans widening because of the inexorable movements of the Earth's crust.

I'll call it plate tectonics

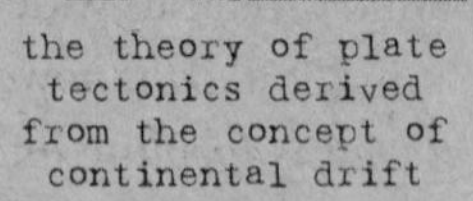

the theory of plate tectonics derived from the concept of continental drift

CONTINENTAL JIGSAW

* *One thing that immediately strikes you about a map of the world is that the continents of South America and Africa look as if they fit together like pieces of a jigsaw.* People had realized this for a long time before the

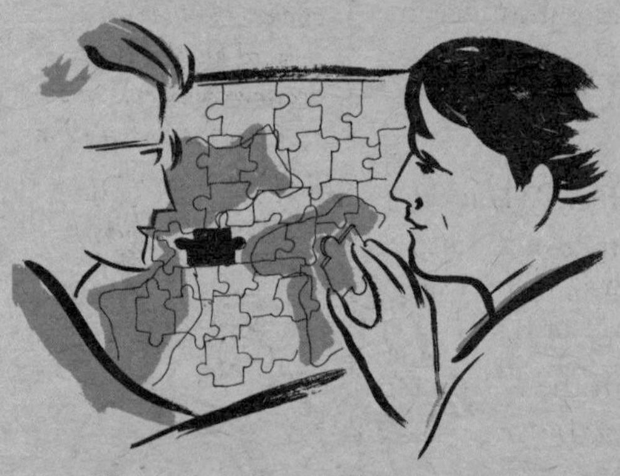

"It must go here! The continents fit together."

German scientist ALFRED WEGENER put forward an explanation in the early 20th century. ***Some 200 million years ago, he suggested, the two continents formed part of a supercontinent (known as Pangaea), and have since drifted apart.***

KEY WORDS

CONTINENTAL DRIFT: the gradual movement of the land masses across the face of the Earth

PANGAEA: a supercontinent made up of all the Earth's land masses joined together

PLATE TECTONICS: the theory that the Earth's surface is made up of a number of segments, or plates, that are slowly moving

* This concept of CONTINENTAL DRIFT gave rise to THE THEORY OF PLATE TECTONICS—*the theory that the Earth's crust is not a solid shell, as was once thought, but is made up of separate* PLATES *of rock that are in constant motion.*

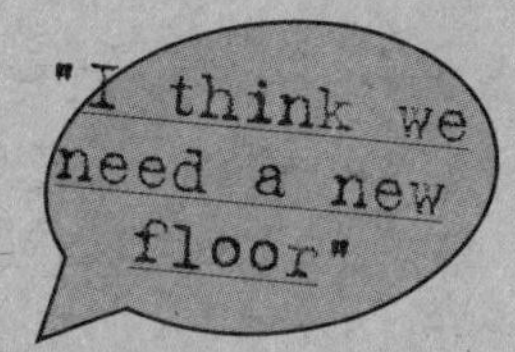

Alfred Wegener (1880–1930)

The son of a Berlin preacher, Alfred Wegener was a meteorologist rather than a geologist, which caused some resentment in geological circles. He first put forward his theory of drifting continents in an article published in 1912.

MOVING PLATES

* *Each* CRUSTAL PLATE *sits on a bed of softer rock, in which heat from the Earth's interior sets up convection currents that carry the plates with them*. South America and Africa sit on separate plates, which are moving in opposite directions, causing the Atlantic Ocean to widen. ***This doesn't leave a hole in the Earth, however, because new plate material is forming on the sea floor, as the two plates separate***. This is called SEA-FLOOR SPREADING, and it occurs under most oceans.

WHEN PLATES COLLIDE

* ***If some plates are moving apart, others must be moving together. When this happens, geological mayhem ensues***. Off the western coast of South America, the South American plate (traveling westward) is in collision with the Nazca plate (traveling eastward) and is forcing the Nazca plate down into the Earth, at some cost. ***It has wrinkled up, to produce the Andes; the friction between the two plates causes earthquakes; and molten Nazca plate material has forced its way up through the crust to create a phalanx of fuming volcanoes.***

when plates crash, it is usually at some cost

THE ROCKY LANDSCAPE

★ Three main kinds of rock are found in the Earth's crust—some born in the fiery cauldrons of volcanoes, others when ancient seas dried up. Once they reach the Earth's surface, they are subjected to relentless attack by the elements.

igneous rocks are born in the fiery cauldrons of volcanoes

granite contains lots of crystals

KEY WORDS

MAGMA: molten rock
IGNEOUS ROCK: rock formed by the solidification of cooling magma
SEDIMENTARY ROCK: rock formed from compressed layers of sediment
METAMORPHIC ROCK: rock that has had its character transformed by heat and pressure underground

FIRE ROCKS

★ *When the* MAGMA *(molten rock) spewed out by volcanoes cools and solidifies, it forms a dark rock called* BASALT*—one kind of* IGNEOUS (FIRE-FORMED) ROCK. When magma solidifies underground, GRANITE is formed. Since the magma cools more slowly there, minerals in granite have time to grow crystals, which gives it a speckly appearance.

LAYERED ROCKS

★ *As time goes by, mud and sand deposited in the sea by rivers build up into layers of sediment, which form* SEDIMENTARY ROCK. Mud layers turn into the rock we call SHALE, sand layers into SANDSTONE. Other types of sedimentary rock, such as LIMESTONE, are formed by layers of chemicals once dissolved in ancient seas. Chemically, CHALK is similar to limestone—but it is made up of skeletons of minute sea creatures.

UNDER THE WEATHER

★ We think of rocks as being hard and everlasting, but over long periods of time they eventually crumble to dust.

★ ***The*** EROSION, ***or wearing away, of the landscape is often caused by the weather***. Sun, frost, ice, rain, and wind all play their part. ***Flowing water is the other principal agent of erosion***. It may attack rocks chemically, dissolving their minerals. ***Limestone is particularly prone to chemical attack, which is why limestone landscapes are often riddled with caves***. But flowing water also mounts a physical

it's not only people that get worn down by the weather

attack. The pebbles and boulders that it carries along grind away the surrounding rocks. ***Some of the finest scenery on Earth has been created in this fashion—such as the Grand Canyon, in Arizona.***

it's crystal clear to me

Changed rocks

Both igneous and sedimentary rocks can be changed (metamorphosed) by heat and pressure within the Earth's interior, giving rise to what are called **metamorphic rocks**. *Slate (once shale) and marble (once limestone) are examples.*

A GRAND CANYON

More than 185 miles long and in places as much as a mile deep, the Grand Canyon in Arizona was created by the Colorado River. The river has been cutting through the landscape for a million years or more, exposing layer after layer of sedimentary rock in a kaleidoscope of different colors and a multiplicity of textures.

LIFE ON EARTH

* The greatest difference between the Earth and the other planets lies in its capacity to support life in abundance. It is the only place in the solar system that has plentiful water, air to breathe, and a suitably warm climate.

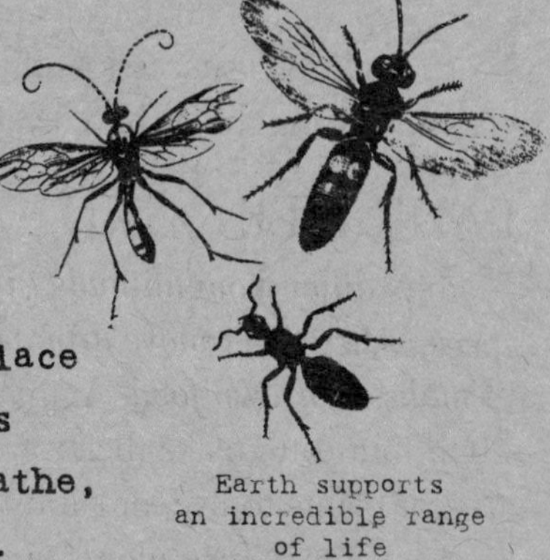

Earth supports an incredible range of life

KINGDOMS OF LIFE

At a conservative estimate, at least 1.5 million different species exist on Earth. Biologists classify them into groups called **kingdoms**, with **plants** and **animals** forming the groups that contain the largest number of species. Simple organisms called **protozoa** are assigned a separate kingdom of their own. So are **bacteria**, and so are **fungi**.

fossils record how life has evolved

plants can make their own food, animals can't

THE FOSSIL RECORD

* *The fossilized remains of bacteria-like organisms tell us that life began on Earth at least 2 billion years ago. But not until about 600 million years ago did a veritable explosion of life take place, as witnessed by the fossil record preserved in the rocks of the*

Cambrian period of the Earth's history. An almost continuous succession of fossils shows how living things have evolved since.

PLANT LIFE

★ ***Plants differ from animals in two main respects. They don't move around, and they can make their own food***. Animals can move around, but can't make their own food; they must eat plants to live, or eat other animals that eat plants. ***Green plants make their food—from carbon dioxide (taken in from the atmosphere) and water (taken in through their roots)—by*** PHOTOSYNTHESIS, ***using the energy in sunlight. Oxygen, which is given out as a waste product into the atmosphere, allows living things to breathe***.

ANIMAL LIFE

★ Animal life on Earth is incredibly varied. The simplest species, like the amoeba, consist of a single cell that carries out all the functions needed to keep them alive. But most animals are MULTICELLULAR CREATURES, ***with specialized body organs that enable them to sense their environment, breathe, feed, and reproduce***. The majority of animal species, such as insects, amphibians, and reptiles, have COLD BLOOD, ***which limits their activity***. Only mammals and birds have WARM BLOOD, ***which allows them to remain active all the time***.

the majority of animal species are cold-blooded

FROM THE SEAS

Early life developed in the seas. There was no life on land until about 400 million years ago. Plants colonized the land first. They were followed by insects and amphibians, which could live on land or in water. Before long, scaly-skinned reptiles became the dominant life form, with the dinosaurs growing to enormous size. When they disappeared off the face of the Earth, about 65 million years ago, mammals began to flourish. Today, one species of mammal—the human species, dubbed *Homo sapiens* (literally, "wise" or "knowledgeable" man)—holds sway over life on Earth.

KEY WORDS

CELL: the smallest unit in the body of living things

PHOTOSYNTHESIS: the process by which green plants make their food

EARTH'S COMPANION

* The moon is the Earth's closest companion in space, and the only heavenly body yet to be visited by the human race. Its regular passage through the heavens has provided a useful clock for humankind for thousands of years.

LUNAR ATTRACTIONS

* The moon is the Earth's only natural satellite. *It is a rocky body, and its diameter is about a quarter that of the Earth. Being much smaller, it has much less mass than the Earth and therefore much weaker gravity—only one sixth that of Earth. Nevertheless, lunar gravity does affect our planet, by causing* TIDES.

* Because it has such weak gravity, the moon hasn't been able to hold onto any atmosphere. *Conditions on the airless moon are harsh. Temperatures soar to more than 212°F during the lunar day, but plummet far below freezing, to lower than –300°F, during the lunar night.*

"Look at the man in the moon, mom!"

KEY WORDS

LUNAR:
to do with the moon (from luna, the Latin for moon)

PHASES:
the changing appearance of the moon in the sky

MOON DATA

Diameter: **2,160 miles**
Average distance from Earth: **240,000 miles**
Mass (Earth = 1): **1/81**
Density (water = 1): **3.3**
Gravity (Earth = 1): **1/6**
Spins on axis in: **27 1/3 days**
Orbits Earth in: **27 1/3 days**
Goes through phases in: **29 1/3 days**

IN A SPIN

★ *Lunar nights and lunar days are both about the same length as two weeks on Earth. This happens because the moon spins around relatively slowly in space—once every 27⅓ days. By a strange coincidence, the moon travels around the Earth once in the same time. The consequence of this is that the moon always presents the same face toward us* (THE NEARSIDE). *The other side* (THE FARSIDE) *is always hidden. This effect is called* CAPTURED ROTATION.

Librations

We can see a little more than half the moon's surface from Earth, because of changes in the tilt of its axis and in its orbital speed. We call these changes **librations**.

the phases of the moon

1 2 3 4 5 6 7

(1) crescent moon (2) half moon or first quarter (3) gibbous moon (4) full moon (5) gibbous moon (6) half moon or last quarter (7) crescent moon. The moon is said to "wax" as more of its face becomes visible, and "wane" as the visible area decreases. It goes through this cycle every 29½ days. This period, with adjustments, forms the basis of our calendar month.

FROM NEW TO FULL

★ *The moon does not emit its own light. It reflects light from the sun. So, as the moon orbits the Earth, we see different parts of it lit up. Our changing views of the moon—its* PHASES*—depend on its position relative to the sun. When it is between the sun and the Earth, the side facing us isn't illuminated, so we can't see it. This is the phase we call the* NEW MOON. As the moon moves on, light peeps around the edge, spreads until half the surface is lit up, then begins to wane.

LUNAR ECLIPSES

The Earth also casts a shadow in space, of course, which the moon enters from time to time. **Such lunar eclipses happen at the time of the full moon. Because the Earth's shadow is quite broad, the moon can stay eclipsed for up to 2½ hours.** But it never completely disappears from view, because it is faintly illuminated by **earthshine**, in other words, sunlight refracted by the Earth's atmosphere.

IN HEAVENLY SHADOWS

* Roughly every 18 months or so, somewhere on Earth awesome moments occur when day turns suddenly into night, the air chills, and birds cease singing and start to roost. This happens on the occasions when a new moon blocks our view of the sun.

SOLAR ECLIPSES

* A TOTAL ECLIPSE of the sun is not only a spectacular event; ***it also provides the chance to observe features of the sun usually obscured by its glare***. Only during a total eclipse can we see prominences leaping out of the chromosphere, and the sun's pearly-white corona (see page 122).

eclipses of the sun and moon

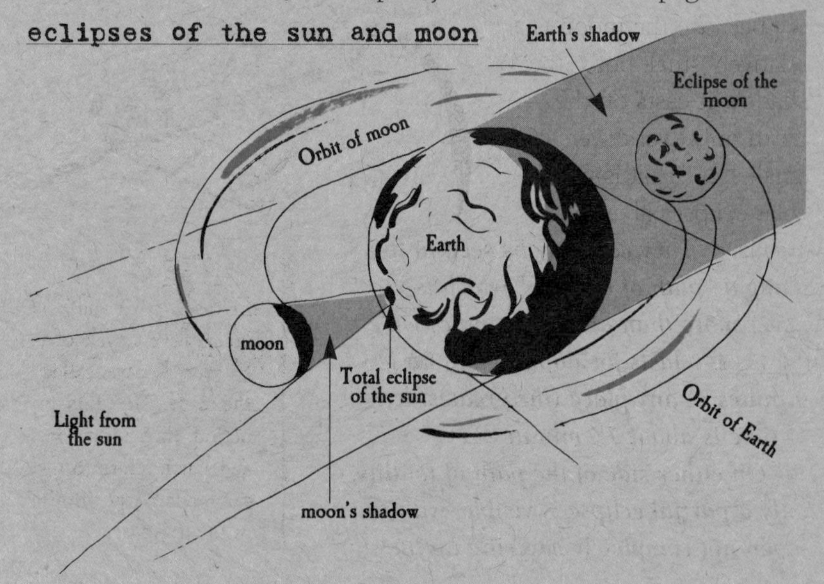

IN SYZYGY

★ *Eclipses occur because of serendipitous celestial geometry. The sun is 400 times farther away from the Earth than the moon is, but its diameter is 400 times greater. The result it that sun and moon appear almost exactly the same size in Earth skies.*

★ Also, the moon circles around the Earth in almost the same plane as the Earth circles around the sun, which is why we see regular phases every month. *We only get eclipses when sun, moon, and Earth are all lined up exactly, or in* SYZYGY *(a handy Scrabble word, derived from the Greek for "yoked together").* And that doesn't happen often.

KEY WORDS

ECLIPSE:
the masking of one heavenly body by another

TOTALITY:
the period of total solar eclipse, during which the sun is completely covered up

THE PATH OF TOTALITY

★ Because the moon is a relatively small body, the shadow it casts on the Earth only ever covers a small area. That is why solar eclipses are so localized. *They can only be seen in full along a "path of totality," which is never more than about 168 miles wide.* TOTALITY *lasts for no more than a few minutes at any place (the greatest length of time is about 7½ minutes).*

★ *On either side of the path of totality, only a partial eclipse is visible, with the moon not completely covering up the sun.*

THE SAROS

Eclipses (solar and lunar) have **a cycle of 18 years 11 days**, called **the saros**. After this period, the sun, moon, and Earth return to the same relative positions in the heavens.

THE LUNAR LANDSCAPE

*** Close up, the surface of the moon displays a host of interesting features—rolling plains, high mountain ranges cut with valleys, snaking ridges, long walls, and deep channels. And there are craters in profusion.**

The volcanic connection

Not all lunar craters have been caused by meteorite impact. Some have a volcanic origin, being the peaks of ancient volcanoes. Volcanic action has also created other surface features, which are particularly evident on the maria. There are smooth raised domes, formed by molten rock pushing upward; and snaking trenches called rilles, which are probably the collapsed remains of underground lava channels called lava tubes (similar ones are found on Earth in Hawaii and other volcanic regions).

the moon is peppered with pits, mountains, and craters

DUSTY SEAS

** With the naked eye, we can see two contrasting features of the moon—light regions and dark ones. Early astronomers thought that the light regions were continents and the dark ones seas (also known as* MARIA). They gave the seas fanciful names—the Sea of Serenity, the Ocean of Storms, the Sea of Nectar. Some regions they thought more sinister, such as the Marsh of Decay and the Lake of Death.

** In fact, the lunar "seas" are not seas but vast, dark plains, created when lava*

see you on the dark side of the moon

flooded into huge impact basins billions of years ago. Most of them are ringed by high mountain ranges, with peaks as much as 20,000 feet high. Among the most prominent ranges are those around the Sea of Showers, which include the lunar Appennines, some 375 miles long. These mountains form part of the lighter-colored highland regions of the moon, called TERRAE, *which are believed to be part of the moon's original crust.*

CRATERS, CRATERS EVERYWHERE

★ *Craters large and small pepper the whole lunar landscape*. They are less common on the maria, however, because they are younger than the highlands. *The craters have mostly been created by the impact of meteorites, which have rained down on the moon since it was formed. There are numerous large craters. The biggest, Bailly, is some 183 miles across*. The largest ones have steep, terraced walls and mountains in the middle of the floor. *At full moon, some craters, such as Tycho, look particularly prominent because of the long glistening streaks called* CRATER RAYS *that emanate from them*.

THE FARSIDE

Although maria dominate the nearside of the moon, they are virtually absent from the farside. This is probably because the farside has a thicker crust, which would have prevented lava from seeping up to the surface to create "seas". Only one small mare (the Sea of Moscow) and a few craters with dark floors, such as Tsiolkovsky, are to be found here. Many of the farside features have Russian names, because a Russian space probe, *Luna 3*, brought us our first view of the farside, in 1959.

KEY WORDS

CRATER:
a pit in the surface, created by meteorite impact or by a volcano

MARE (PLURAL MARIA):
a lunar "sea" (mare being Latin for sea)

TERRA (PLURAL TERRAE):
highland region (terra being Latin for land)

Those famous first words

The first moon landing, and Neil Armstrong's famous words— ***"That's one small step for a man, one giant leap for mankind."*** *—were seen and heard by over half a billion people back home on Earth. The astronauts spent over two hours on the moon's surface and returned to a heroes' welcome after eight days in space.*

"That's one small step for a man, one giant leap for mankind!"

THE APOLLO ASSAULTS

★ Between July 1969 and December 1972, Apollo astronauts took a "giant leap for mankind" by exploring the moon on foot. It is thanks to them that we now know what the moon is really like.

KENNEDY'S COMMITMENT

★ In a speech before Congress on May 25, 1961, President John F. Kennedy exhorted the U.S. to reach for the moon, with the historic statement: ***"I believe that this nation should commit itself to achieving the goal, before this decade is out, of landing a man on the moon and returning him safely to Earth."***

★ And, sure enough, before the decade was out, the Apollo moon-landing project was ready to deliver the greatest spectacle in history—***humans walking on another world***.

THE BUILD-UP

★ At the time of Kennedy's speech, a moon-landing seemed an impossible dream. ***But, within eight years, successful orbital flights by the Mercury astronauts (who went into space one by one) and Gemini astronauts (who went in two by two) showed that humans could survive and function well in space.***

the Saturn V rocket had as much power as...

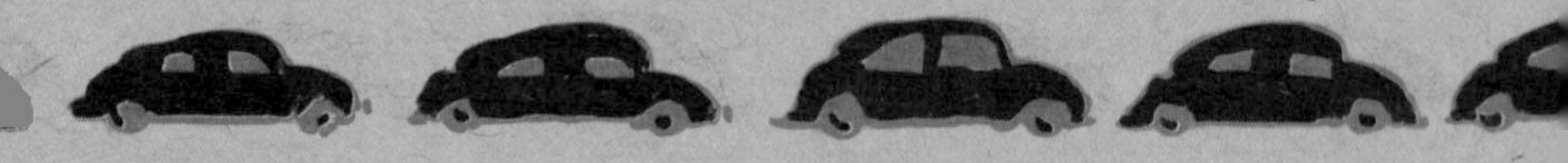

★ While the astronauts were honing their skills, gargantuan hardware was being designed and built. ***Most impressive was the* Saturn V *rocket, 36 stories high and weighing 3,000 tons. Its power, as NASA said at the time, was equivalent to that of "a string of Volkswagens" stretching from New York to Seattle***. It needed such power to boost a three-part Apollo spacecraft into a trajectory toward the moon.

Kennedy urged NASA to reach for the moon

THE EAGLE LANDS

★ At last, on July 16, 1969, three astronauts blasted off from the Kennedy Space Center in Florida, bound for the moon. ***Four days later, Neil Armstrong and Buzz Aldrin touched down on the moon's Sea of Tranquility in their lunar module, call sign "Eagle." Just a few hours later, Armstrong planted the first human footprint on lunar soil***.

EXPLORING SEAS AND HIGHLANDS

Armstrong and Aldrin were the first of a dozen astronauts who explored the moon's surface in six different locations—on the maria ("seas") and in the highlands. They collected rock and core samples, carried out experiments, and set up a series of automatic scientific stations that would radio measurements back to Earth after they had left. And they took thousands of photographs, which still take the breath away, even today.

...a string of Volkswagens stretching from New York to Seattle

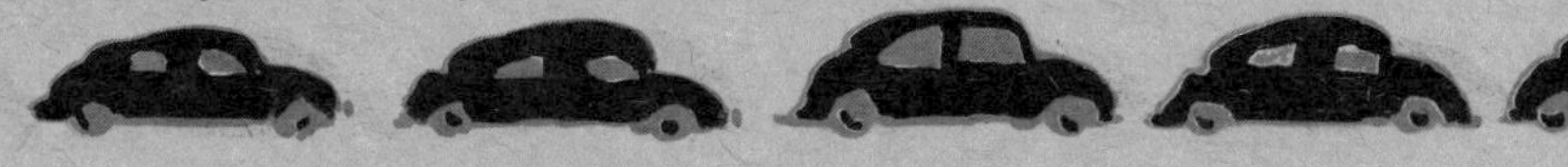

THE ASTRONAUT'S MOON

* From the rock and soil samples brought back by the astronauts, the measurements relayed by the scientific stations, and the data gathered by the orbiting mother ships, we now know that the moon is not made of green cheese.

geologists have studied moon rock samples

oh no...

sadly for mice, the moon proved not to be made of cheese

Moon rocks

The rock typical of the lunar seas is a dark basalt—not unlike the basalt found on Earth, formed when lava flows solidify, creating igneous rocks. The typical rock of the moon's highlands is a lighter-colored breccia. This is a kind of rock pudding consisting of a mixture of fragments of older rock. It is formed when rocky material that has been made molten by the force of a meteorite impact cements existing rock chips together.

THE DUSTY SURFACE

* It was feared that the surface dust on the moon might be so deep that it would swallow any landing craft. Happily, this fear proved to be unfounded. ***There is dust, but only a couple of inches thick. It forms part of a lunar soil, known as*** REGOLITH, ***which has the consistency of newly plowed soil on Earth. The soil has been produced by the constant bombardment of meteorites, which pulverize the surface rocks as they dig out craters. The soil particles include masses of tiny glass beads, which make the ground slippery to walk on, as the astronauts soon found out.***

VOLCANIC ROCKS

* On the surface of the moon, rocks are strewn everywhere. They are all volcanic rocks. There are no sedimentary rocks, as

there are on Earth. The Earth's sedimentary rocks formed from layers of sediment that settled out of ancient seas. ***And since there have never been any seas (in the ordinary sense of the word) on the moon, there can't be any sedimentary rocks.***

the moon's structure

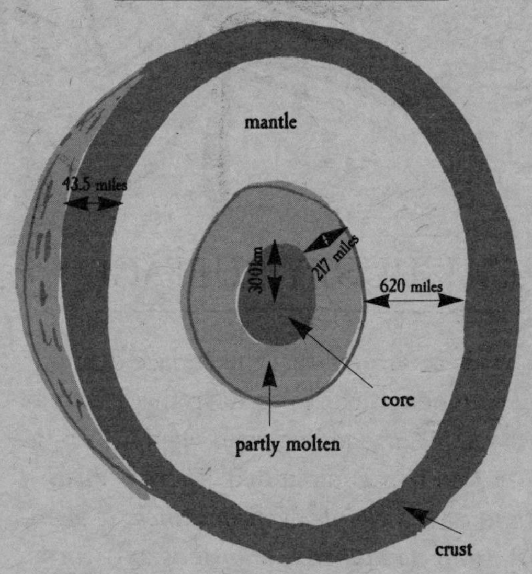

from seismometer readings sent back over several years, we now know that the moon's structure is broadly similar to the Earth's

MOON STRUCTURE

★ Among the instruments that the Apollo astronauts set up on the moon were SEISMOMETERS, to measure tremors. Here, on Earth, geologists use similar instruments to pick up earthquake waves, which helps them to study the structure of the Earth's interior and forecast earthquakes. ***On the moon, seismometers were used to detect "moonquakes" and trace how seismic waves travel through the interior.***

EARTH'S DAUGHTER NO MORE?

The Apollo results show that the moon is chemically different from the Earth, which would appear to discount the once widely held belief that it was torn from the Earth. **The theory favored at present is that the moon was formed largely from the remnants of a huge body that struck the very young Earth.**

Interior structure

From seismometer readings, scientists now have a good idea of the interior structure of the moon, which is layered like the Earth's. The outermost layer of the crust seems to be made up of cracked and broken rock, with more solid rock underneath. Below the crust comes a thick mantle, composed of another kind of rock, the bottom of which may be molten. Then at the center of the moon is a small solid core, probably containing iron, like the Earth's.

RED PLANET MARS

★ Known as the Red Planet because of the fiery reddish-orange color we see shining from it in the night sky, Mars is named after the Roman god of war. Similar to the Earth in some ways, it is the only planet where human beings could possibly set up home—but they would need to take everything with them, including oxygen and water.

Mars, the Roman god of war

OPPOSITION

From Earth, Mars is best seen when it lies exactly opposite the sun in the sky, as do all the other outer planets. In the case of Mars, oppositions come around about every 26 months. But some oppositions are better than others, because of the planet's somewhat eccentric orbit around the sun.

KEY WORDS

OPPOSITION: the position of an outer planet exactly opposite the sun in the sky

MARS IN THE SKY

★ *Mars is one of the smallest planets, with a diameter about a quarter that of the Earth*. But at OPPOSITION (when it is opposite the sun in the sky), it can shine as bright as the giant planet Jupiter, because it is very much closer to us. ***At its closest, it comes within 35 million miles, closer than any planet except Venus***.

"You'll have to delay dinner, there's going to be an opposition"

★ Telescopic examination reveals various dark markings, particularly in a region called Syrtis Major near the Martian equator, and icecaps at the north and south poles. ***The icecaps come and go seasonally. Like Earth, Mars has seasons, because its axis tilts.***

ANOTHER EARTH?

★ Mars also has an atmosphere, and a day that's only fractionally longer than our own.

Mars is the only planet where human beings could set up home

But it could never be another Earth. Its atmosphere is much slighter than Earth's, and the main gas in it is carbon dioxide.

★ ***Mars is also a much colder planet than our own.*** Temperatures only just struggle above freezing, even at noon, in the Martian summer. At nights, with not much of an atmosphere to act as a protective blanket, temperatures plunge to –250°F or below.

MARS DATA

Diameter at equator: **4,216 miles**
Average distance from sun: **141,000,000 miles**
Density (water=1): **3.9**
Spins on axis in: **24 hrs 37 mins**
Circles sun in: **687 days**
Moons: **2**

MARTIAN MOONS

The American astronomer **Asaph Hall** (1829 –1907) discovered Mars's two tiny moons in 1877. Named **Phobos** and **Deimos**—after the two horses that pulled Mars's war chariot in Roman mythology—they are truly tiny. Phobos is about 17 miles across, and Deimos only about 10 miles. Astronomers reckon that they are probably asteroids that strayed from the asteroid belt (see page 170), which is relatively near, and were captured by Mars's gravity.

THE MARTIAN LANDSCAPE

* When space probes began orbiting and landing on Mars in the 1970s, they revealed why the planet appears red in the sky. Its surface is covered with rust-red soil and rocks. They also sent back pictures of a fascinating landscape, dominated by a quartet of huge extinct volcanoes and a Martian "Grand Canyon" capable of engulfing the famous terrestrial one many times over.

ICE AND WATER

Space probes have revealed yet another important feature that Mars and Earth have in common—water. They have sent back pictures of water-ice clouds clinging to Martian mountain-sides and filling the planet's deep canyons. They have also found that ordinary ice accounts in part for the icecaps, along with what we call "dry ice" (frozen carbon dioxide). Some of the landscape features indicate that there was once flowing water on Mars, though many millions of years ago.

plains, craters, canyons, mountains, volcanoes—Mars has it all

BIRD'S-EYE VIEW

* *Mars is a planet of contrasts. Some regions are quite heavily cratered, due to repeated bombardment by meteorites long ago. The southern hemisphere has two great basins formed by massive impacts.* The one, Hellas, is more than 1,000 miles across—about twice the size of the other, Argyre. During the southern winter, frost and mist often hang around these basins.

MARTIAN MOUNTAINS

★ The highest parts of Mars occur in the region known as THARSIS, just north of Mars's equator. ***It is there we find four of the biggest volcanoes in the solar system. By far the biggest is*** OLYMPUS MONS ***(named after Mount Olympus, home of the gods in Greek mythology)***. This massive mountain stands some 16 miles high—three times the height of Mount Everest. At its base, it is 375 miles across.

THE GRAND CANYON

★ ***The most extensive feature on Mars begins not far from the Tharsis Ridge and runs close to the equator for some 3,100 miles***. Named VALLES MARINERIS (MARINER VALLEY), after the *Mariner* 9 space probe that discovered it early in 1972, it was immediately dubbed the Martian "Grand Canyon." ***A great gash in the Martian crust, it has an average depth of 4 miles and in places is as much as 250 miles wide***. Lesser canyon systems fan out from it in all directions, making it the most gigantic geological feature we know anywhere.

ON THE PLAINS

★ ***Vast, flat plains*** (PLANITIA) ***are also a feature of the Martian landscape. They are dusty deserts strewn with rocks.*** Even though the Martian atmosphere is slight, it still allows very strong winds to blow, which kick up the dust. ***Sometimes these storms extend over most of the planet***.

THE MARTIAN ROVER

In 1997 the *Pathfinder* probe carried to Mars a tiny rover called *Sojourner*, about the size of a microwave oven. For more than two months *Sojourner* analyzed nearby rocks. It found the expected volcanic rocks, such as basalt, but also pictured what could be **conglomerates**. These are sedimentary rocks—rocks that would have been formed as a result of action and deposition by rivers.

"But there's a fundamental difference!"

A crucial difference

There's a fundamental difference between Mariner Valley and the real Grand Canyon. The latter was formed by erosion by the mighty Colorado River. Mariner Valley is a geological fault, formed when the crust collapsed.

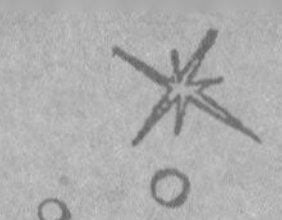

LIFE ON MARS

* Of the nine planets in the solar system, Mars has always been considered the one most likely to harbor life of some kind, in addition to Earth. A century ago some astronomers were convinced of it. Today, we know that Mars is far too hostile a world for life to exist there now. But it's possible that there was life on Mars in the distant past.

No, I said "channels"

Schiaparelli didn't mean "canals"

Percival Lowell (1855–1916)

US astronomer Percival Lowell was one of the most adamant believers in the existence of intelligent life on Mars. In 1894 he built an observatory at Flagstaff, Arizona, for the specific purpose of studying the planet. Its fine 24-inch telescope is still in use today.

THOSE MARTIAN CANALS

* The idea of life on Mars took off after a report by an Italian astronomer, GIOVANNI SCHIAPARELLI (1835–1910), that he had seen "canali" on Mars. He meant "channels"—but the Italian word was misinterpreted as "canals," implying waterways built by intelligent beings.

* Astronomers also noticed a "wave of darkening," that looked like vegetation, sweeping across the planet in the spring. ***And some envisaged a dying Martian race channeling water from the melting icecaps to irrigate crops in farming regions.***

MARTIANS HAVE LANDED!

* On the evening of October 30, 1938, listeners to CBS radio had their programs interrupted by a news flash. ***An object had fallen from the sky in New Jersey, said the announcer, that was not a meteorite and contained "strange beings who are believed to be the vanguard of an army from the planet Mars."***

★ Panic rushed through a good proportion of the listeners who didn't stay tuned to the end—when they would have learned that this was a radio adaptation of H. G. Wells's *War of the Worlds*, made vividly convincing by the young Orson Welles.

SPACE-AGE MARS

★ When space probes revealed that Mars was a nearly airless desert, the prospects of life there fell to zero. Then, when the Viking probes landed on the planet, they tested the soil for signs of life, but in vain. ***Nevertheless, the* Viking *and later the* Pathfinder *and* Sojourner *probes found plenty of evidence that water once flooded across Mars. This suggests that conditions were once more hospitable and might have allowed some kind of life to thrive.***

Martians?

On the radio?

ALH84001

A new twist in the life-on-Mars debate surfaced in 1996, when scientists at NASA claimed that a Martian meteorite discovered in the Antarctic—known as ALH84001—contained microfossils of some kind of primitive bacteria. Although other planetary scientists have disputed this finding, one can be sure that few of them will be surprised if fossils do show up when samples of Martian soil are returned from Mars in a few years' time.

KING OF THE PLANETS

*** Twice as massive as all the other planets of the solar system put together, Jupiter is named after the king of the gods and ruler of the heavens in Roman mythology. It is also very different from the planets we have looked at so far, which are mainly rocky. Jupiter is mainly gassy.**

Jupiter, king of the heavens

Amazing spin

Jupiter is more than three times farther away from the sun than Mars, and takes nearly 12 years to circle the sun. Huge though it is, it spins on its axis at amazing speed, taking only about ten hours to spin around once. As a result, it has the shortest "day," or period of rotation, of all the planets.

JUPITER DATA
Diameter at equator: **88,350 miles**
Average distance from sun: **483,000,000 miles**
Density (water=1): **1.3**
Spins on axis in: **9 hrs 50 mins**
Circles sun in: **11.9 years**
Moons: **16+**

JUPITER IN THE SKY

* ***We can see Jupiter clearly with the naked eye***. At opposition (see page 150), which occurs roughly every 13 months, it reaches a magnitude of up to –2.5, outshining all the stars and rivaling Mars. ***But it can easily be distinguished from Mars because it is a brilliant white, while Mars is definitely orange in color.***

THROUGH A TELESCOPE

* A telescope reveals a host of fascinating features on a colorful disk, which isn't the planet's surface but its thick atmosphere. ***Overall, it presents a striped appearance, with alternating light and dark bands, which astronomers call the zones and belts. They are light and dark clouds stretched out parallel by the planet's rapid rotation***. The swift rotation also makes Jupiter noticeably OBLATE, or flattened at the poles.

SPOTTED PLANET

* Within and around the belts and zones there are all kinds of spots, whorls, and wisps. These reflect the frenetic eddying and turbulence set up as the clouds surge through the atmosphere. ***Most features come and go over periods of days or months. But one is virtually permanent, having been visible for most of the past three centuries. It is well named the*** GREAT RED SPOT.

GREAT RED SPOT

* The English physicist ROBERT HOOKE (1635–1703) first observed the Great Red Spot in 1664. ***It is an oval region, just south of Jupiter's equator, up to 25,000 miles long. Space probes have shown it to be the eye of an enormous storm in which clouds swirl furiously in a counterclockwise motion. The redness seems to be due to large amounts of phosphorus.***

THE LAYERED STRUCTURE

Jupiter's atmosphere is at least 600 miles thick. It is made up mainly of hydrogen and helium (as is the sun). The clouds seem to be made up of ammonia crystals and ammonium compounds with sulfur. Beneath the atmosphere lies a deep ocean of hydrogen in a liquid form. **At the bottom of the ocean, the pressure is so incredibly high that the hydrogen is forced into a very dense metal-like state called metallic hydrogen. The currents circulating in this layer give Jupiter a magnetic field—like the Earth's, but much stronger (see page 133).** At the very center of the planet lies a relatively small ball of rock.

JOVIAN MOONS

* Like the other giant planets, Jupiter has a veritable swarm of moons circling around it. Some are only tiny and have far-flung orbits, but others are as big as planets and circle close in. The four largest ones can even be seen with binoculars.

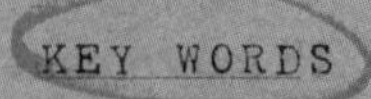
KEY WORDS

GALILEAN MOONS: Jupiter's four biggest moons (Io, Europa, Ganymede, and Callisto), first spotted by Galileo

JOVIAN MOONS: the moons belonging to Jupiter

tiny Metis and Adrastea probably act as shepherds

A JOVIAN RING

Another significant *Voyager* discovery was the presence of a ring around Jupiter. But it bears no comparison with Saturn's rings: it is very faint and made up of fine particles. Two tiny moons, Metis and Adrastea, move in the same region and probably serve as "shepherds" (see page 163).

THE GALILEAN MOONS

* Jupiter's four biggest moons were discovered by Galileo in the winter of 1609–10, when the great man was feasting his eyes on the heavens with a telescope for the very first time. ***He saw them, as we can today in binoculars, as tiny points of light moving in line from one side of Jupiter to the other.***

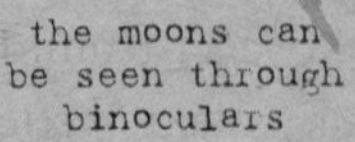
the moons can be seen through binoculars

* ***In order of increasing distance from their parent planet, the so-called*** GALILEAN MOONS ***are Io, Europa, Ganymede, and Callisto.*** The largest of the four are Callisto and Ganymede. ***Callisto is only fractionally smaller than the planet Mercury. And, with a diameter of 3,250 miles, Ganymede is not only larger than Mercury but is the biggest moon in the whole solar system.***

phew, it's pure sulfur!

ICY WORLDS

★ ***All four moons look quite different from one another***. Europa, which is about the same size as our moon, has a very smooth surface. ***In fact, it seems to be the smoothest body in the solar system***. It is probably made up mainly of rock but with a thick coating of ice on top, which makes it highly reflective. Callisto and Ganymede seem to be made of less rock and more ice. Both moons are cratered, especially Callisto; ***and bright spots on their surface show where meteorites have recently crashed on them and gouged out fresh ice***.

A sulfur surface

Io's volcanoes don't spew out molten rock, as volcanoes do on the Earth. They pour out streams of molten sulfur. This provides an explanation of the moon's vivid coloration, since different forms of sulfur range from yellow to deep orange in color.

THE PIZZA MOON

★ ***Io stands out among the Jovian moons in more ways than one. It has the most extraordinary deep orange color, with yellow, white, and black patches***. When *Voyager 1*'s cameras flashed back the first pictures of it in February 1979, the leader of the Voyager team, Bradford Smith, said it looked ***"better than a load of pizzas"*** he'd seen. So it is now often called the pizza moon. ***Shortly afterward, the team spotted an even more unusual feature on Io—active volcanoes, which erupt regularly because of the tidal forces set up within Io by the gravity of its parent planet***.

LORD OF THE RINGS

★ Saturn's trio of shining rings makes the second-largest planet one of the finest sights in the heavens for the telescopic observer. It is also the planet with the largest number of known moons, one of which is as big as Mercury.

scythe-bearing Saturn was probably an agricultural god

SATURN IN THE SKY

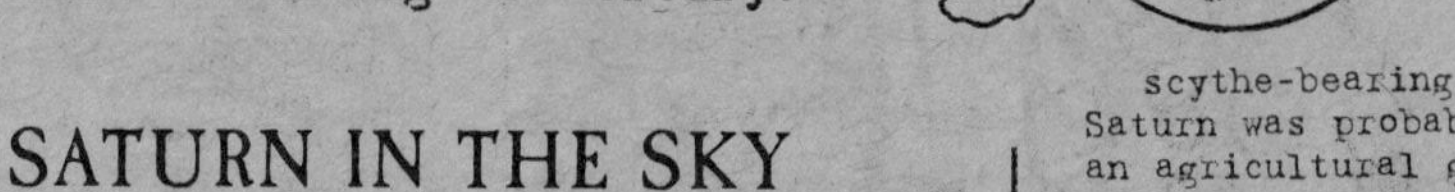

★ *Although Saturn is nearly as big as Jupiter and, like Jupiter, has a highly reflective covering of cloud, it is much less brilliant*. The reason is that it lies twice as far away from the sun. *But on occasions, at opposition (see page 150) it reaches a magnitude of about –0.3, which makes it more brilliant than any star, except Sirius and Canopus. However, for much of the time it is much dimmer and may be difficult to spot*.

IN THE TELESCOPE

★ *Saturn may not seem very impressive to the naked eye or even when viewed with binoculars, but through a telescope it looks truly spectacular*. In 1610 Galileo noticed how different this planet looked from the others and thought it might be a triple-planet system, since the projection of the rings on each side of it looked like

SATURN DATA

Diameter at equator: **74,000 miles**
Average distance from sun: **890,000,000 miles**
Density (water=1): **0.7**
Spins on axis in: **10 hrs 39 mins**
Circles sun in: **29½ years**
Moons: **25+**

Saturn's lights

Since the early 1990s, the Hubble Space Telescope has been keeping a close eye on Saturn. In 1997 it sent back spectacular images showing intense auroras–like the Earth's northern and southern lights–around Saturn's polar regions.

small bodies. It *was left to* <u>CHRISTIAAN HUYGENS</u> *to discover the true nature of Saturn in 1655—though he thought it was encircled by just one ring, because he could only see the system as a single ring in the telescopes he had available*.

Huygens invented the pendulum clock

PALE IMITATION

* *Investigation by space probes has revealed that Saturn is similar in composition and structure to Jupiter*. It has an atmosphere, mainly of hydrogen and helium, above deep layers of liquid and metallic hydrogen, and a central rocky core. It spins around only fractionally slower than Jupiter, and it too is flattened at the poles.

* *Usually the most prominent feature of Saturn's disk is the shadow cast on it by the rings. The planet has cloud bands similar to Jupiter's, but nowhere near as obvious*. There are also spots and other atmospheric disturbances that indicate stormy weather on the planet.

Saturn afloat

Since Saturn's density is only 0.7, if you could drop it into a big enough bowl of water, it would float.

<u>CHRISTIAAN HUYGENS (1629-95)</u>

Huygens was a versatile, practical man and a skilled astronomical observer. He was an expert telescope-maker, who invented the Huygens eyepiece, still widely used. He also improved the design of watches and built the first pendulum clock, which revolutionized timekeeping.

hmmm, hydrogen, helium...

Saturn's composition is akin to Jupiter's

SATURN'S RINGS AND MOONS

★ The other giant planets have their rings, but only Saturn's glorious ring system can be seen from Earth. Although we can only make out three rings from here, there are in fact many more, as space probes have revealed. Some of the rings are attended by tiny moons playing gravitational tag.

Saturn and its rings

MANY MOONS

The tiny shepherds were only two of the new moons space probes have discovered around Saturn, taking its tally of moons from the ten visible in telescopes to the 25 or so we know now. **Christiaan Huygens—the discoverer of Saturn's rings—also discovered the first of its moons, Titan.** That this one should have been the first to be discovered is not surprising, since **it is bigger than the planet Mercury and is the second largest moon in the whole solar system.**

ABOUT THE RINGS

★ *Saturn's rings girdle the planet's equator and have an overall diameter of some 169,000 miles. Of the three that we can make out from Earth, the outer one is referred to as A, the central one as B, and the inner one as C.* The A and B rings are by far the brightest and are separated by the dark CASSINI DIVISION, named after the Italian astronomer GIOVANNI CASSINI who discovered it in 1675. The inner C ring—sometimes called THE CREPE RING—is faint and almost transparent.

NOW YOU SEE THEM...

★ Because Saturn's axis (like the Earth's) is tilted with respect to its orbital path, we see different aspects of the rings during the planet's journey around the sun, which takes nearly 30 years to complete. *At times the rings look wide open; at others they present themselves edge-on, as they did in 1995, and virtually disappear from view. This shows how thin the rings are—in places they are only a few hundred yards thick.*

ROCKS AND RINGLETS

★ ***What exactly are the rings?*** Although they look very solid in telescopes based on Earth, they can't be—because the huge gravitational forces exerted by Saturn wouldn't allow a solid ring to exist. ***They are in fact made up of particles and lumps of ice, dust, and rock, which whiz around at high speed, reflecting sunlight as they do so, giving the overall effect of shining sheets. When viewed in close-up by the*** Voyager ***probes, the three main rings were seen to be made up of literally hundreds of individual ringlets, making the ring system look like a huge celestial long-playing record.***

Saturn's shepherd moons keep its rings in order

RINGS AND SHEPHERDS

★ The Voyager probes discovered several more rings, outside the A ring and inside the C ring. ***They also spotted a pair of tiny satellites at the edge of ring A, seemingly keeping the particles in the ring in place.*** These are now known as shepherd moons.

TITAN

Titan is one of the most interesting of all moons because it has a dense orange-colored atmosphere. **The main gas present is nitrogen, with traces of argon and hydrocarbons such as methane. Since Titan's temperature is close to the boiling point of methane (about –350 °F), it is possible that this substance plays the same role on Titan that water does on Earth.** So there may be methane clouds condensing into methane rain and snow, which fall onto a landscape featuring methane lakes and rivers.

KEY WORDS

SHEPHERD MOONS: moons located around the edge of planetary rings that keep the ring particles in place, like a shepherd herding sheep

TOPSY-TURVY URANUS

* Uranus was the first of the "new" planets to be discovered, being first seen by William Herschel in 1781. It is notable for the excessive tilt of its axis, which effectively means it spins on its side as it orbits the sun. Since most planets spin nearly upright as they orbit, Uranus must have suffered a violent collision with another large body soon after it formed, nearly 5 billion years ago.

Uranus's moons

The ten new moons spotted by Voyager 2 *were all given Shakespearean names (such as Cordelia and Ophelia for the inner-most ones), in line with Uranus's five earlier-known moons (such as Ariel and Miranda).* ***In close-up, all five big moons are interesting. Miranda's surface displays the most incredible geology of any body in the solar system. Vastly different kinds of terrain abut each other—rolling cratered plains suddenly giving way to regions that are spectacularly grooved.***

THE GREEN GIANT

* *The ancients didn't know about Uranus, because even at its brightest it is right on the limit of naked-eye visibility*. Although its diameter is about four times the Earth's, it is so far away that even in binoculars it looks like an ordinary star. *Powerful telescopes show it as a small bluish-green disk but reveal no features. Close-up pictures taken in 1986 by* Voyager 2 *showed a very bland surface*.

the Greek god Uranus governed the universe

THE INSIDE STORY

* Like Jupiter and Saturn, Uranus has an atmosphere composed mainly of hydrogen and helium. ***But it has significant amounts of methane, too***. Uranus has a much more extensive atmosphere than Jupiter and

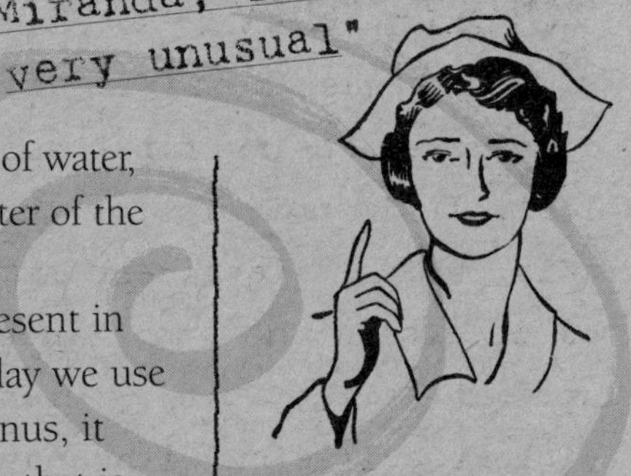

Saturn—***it is probably more than 6,000 miles deep***. Below it there is a deep ocean made up of water, ammonia, and methane. The center of the planet consists of a rocky core.

* Methane gas was probably present in the Earth's early atmosphere. Today we use it for fuel, as natural gas. On Uranus, it is the methane in the atmosphere that is responsible for the planet's color. ***The methane gas absorbs the red wavelengths from white light, leaving the characteristic blue-green hue***.

* ***Uranus became the second planet known to have rings around it when they were discovered by chance in 1977***. They are too faint to be visible in telescopes. *Voyager* 2 revealed that there are at least ten rings, some with tiny shepherd moons (see page 163) on either side. ***Uranus's moons are very narrow—down to as little as 1 mile wide. And they look very dark. Astronomers think they may be bits of rock covered with dark organic compounds***.

Uranus orbits spinning on its side, and shepherd moons keep its rings in order

Remember Miranda

Astronomers reckon the highly unusual features on Miranda's surface could be the result of a catastrophic collision between a moon and a huge asteroid that occurred millions of years ago. It would have smashed both to pieces, and then gravity would have pulled the pieces back together, leaving the kind of haphazard landscape we see today.

URANUS DATA

Diameter at equator: **32,000 miles**
Average distance from sun: **1,780,000,000 miles**
Density (water=1): **1.3**
Spins on axis in: **17 hrs 14 mins**
Circles sun in: **84 years**
Moons: **17+**

THE BLUE PLANET, NEPTUNE

★ Neptune was unknown until 1846, when it was discovered by German astronomer Johann Galle. A near twin of Uranus in size and makeup, it has much more weather in its lovely blue atmosphere, where furious winds blow and storms rage. It also has a ring system—more prominent than the one around Uranus, but impossible to see from Earth.

Neptune, the trident-bearing Roman god of the sea

DISTANT TWIN

★ *Neptune is only fractionally smaller than Uranus, but lies nearly a billion miles farther away. It is a tiny object when seen through a telescope, showing just two moons*. Practically all our knowledge about it has come from spacecraft—from *Voyager* 2 in 1989, and more recently from the Hubble Space Telescope.

★ Neptune is nearly identical to Uranus in composition. A very thick atmosphere of hydrogen, helium, and methane overlays a deep ocean of water, ammonia and methane, and below that is a rocky core.

OTHER WAY AROUND

★ As with all the giant planets, the atmosphere on Neptune circles around the planet at high speed. ***The strange thing is, though, that it travels in the opposite direction to the planet's rotation***.

NEPTUNE DATA

Diameter at equator: **31,000 miles**
Average distance from sun: **2,793,000,000 miles**
Density (water=1): **1.8**
Spins on axis in: **19 hrs 12 mins**
Circles sun in: **165 years**
Moons: **8**

More rings and more shepherds

Neptune has four faint rings—two broad and two narrow. Tiny moons, both inside and outside the rings, shepherd the ring particles.

HEAT FROM WITHIN

★ *Even though Neptune lies much farther away from the sun than its twin, it is about the same temperature, if not a little warmer*. The temperature at the cloud tops is about –420°F. *This must mean it has some kind of internal heat source, so that its oceans are relatively warm*.

NEPTUNE'S WEATHER

★ It is probably the warm oceans that make Neptune's weather so different from that of Uranus. Bands of white clouds, made up of methane-ice crystals, form in many regions. *Voyager 2 spied a swift-moving cloud formation, nicknamed* THE SCOOTER *because it circled the planet so fast; and a large, dark storm center,* THE GREAT DARK SPOT. *Around it, the most violent winds we know in the solar system whirl at 1,250 mph.*

ICY GEYSERS

Two of Neptune's moons are visible from Earth—Triton and Nereid. **Triton is the biggest and by far the most interesting. For one thing, its temperature is –455°F, making it one of the coldest bodies we know. For another, it shows volcanic activity, which seems incongruous for such an icy world.** But we are not speaking of molten-rock volcanoes, like Earth's. **Triton's volcanism—at least the kind spied by *Voyager 2*—takes the form of geyserlike activity. Vents spew out plumes of nitrogen gas full of black dust, which settles and streaks the landscape.** The overall color of Triton is pinkish. This is probably due to deposits of methane ice, colored by organic compounds.

PLUTO AND CHARON

★ Pluto was the last planet to be discovered, by the American astronomer Clyde Tombaugh in 1930. It is a tiny deep-frozen world only two-thirds the size of the moon. Remarkably, its own moon Charon is half its size, so we should probably regard Pluto-Charon as a double planet.

A TINY WORLD

★ *Pluto is the tiniest planet in the solar system. Its diameter is less than one-fifth of the Earth's. Its orbit takes it farther away from the sun than any other planet—until it lies nearly 4,600 million miles away from*

CHARON

Pluto's satellite Charon was discovered in 1978 in a photograph in which Pluto appeared elongated. The Hubble Space Telescope has since taken pictures showing the two bodies widely separated. **For a moon Charon is huge, compared with its "parent planet." It also circles very close in, so that it is sometimes only 12,500 miles from Pluto. Another unusual facet of their relationship is that Charon orbits Pluto in 6 days 9 hours—and, since it takes Pluto the same time to spin around, Charon remains over the same spot on Pluto all the time.**

"Did you hear about this small new planet?"

the sun. Being so small and so far away, it is hardly surprising that it was discovered so recently.

ECCENTRIC PLUTO

★ ***Pluto's orbit around the sun is the strangest of any planet. It is the most eccentric (oval) of them all, at times bringing the planet as close as 2,750 million miles to the sun***. And for 20 years of its 248-year orbit around the sun, it circles within the orbit of Neptune—as it did between 1979 and 1999. ***Its orbit also takes it farther away from the general plane of the solar system than any other planet.***

PLUTO'S MAKEUP

★ ***Pluto is a desperately cold world, with a surface temperature of about –450°F. From its density, astronomers reckon it is made up of mainly rock and water ice***. They have also detected traces of methane. No one knows for certain, but Pluto may well have a thin methane atmosphere. On the surface, there is probably a very thick layer of frozen methane and water ice. Then below that, more water ice, forming a mantle around a relatively large rocky core.

on Pluto it's icy cold

PLUTO DATA
Diameter at equator: **1,500 miles**
Average distance from sun: **3,670,000,000 miles**
Density (water=1): **2**
Spins on axis in: **6 days 9 hours**
Circles sun in: **247.7 years**
Moons: **1** (Charon)

Planet X

Pluto's discovery was due to astronomers looking for a planet that would explain the irregularities in the orbits of Neptune and Uranus. But Pluto and Charon have too small a mass to cause them. So, the hunt for a tenth planet, Planet X, went on. ***With the rapid advances in telescope technology, Planet X—if it exists—is going to be found soon.***

CHAPTER 7

BITS AND PIECES

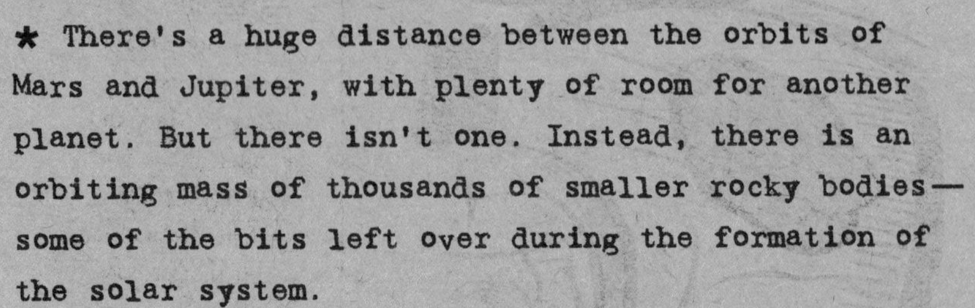

★ There's a huge distance between the orbits of Mars and Jupiter, with plenty of room for another planet. But there isn't one. Instead, there is an orbiting mass of thousands of smaller rocky bodies—some of the bits left over during the formation of the solar system.

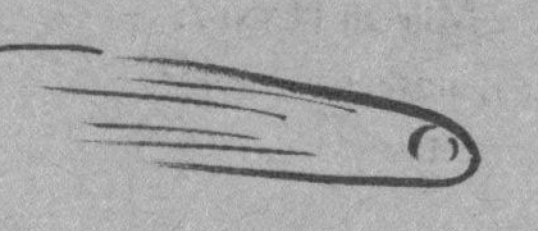

Spotting asteroids

With the exception of Vesta, none of the asteroids can be seen in the night sky with the naked eye. They are too small and too far away. Even Vesta can only just be seen—at its closest approach, in perfect conditions, and by observers with exceptional eyesight.

TINY WORLDS

★ *These small rocky bodies are known as* THE MINOR PLANETS, *or asteroids. An Italian astronomer,* GIUSEPPE PIAZZI, *(1746–1826) discovered Ceres, the first and largest, on January 1, 1801. By 1807 three more had been found—Pallas, Juno, and Vesta. Today, we know of thousands.*

★ Ceres is about 600 miles across. And only two other asteroids, Pallas and Vesta, measure more than 310 miles across. *Most of them are in the tens of miles range.*

THE ASTEROID BELT

★ *Most of the asteroids orbit around the sun in a broad belt, about 93 million miles wide, but quite a number orbit well outside the belt.* Some travel outward way beyond Jupiter's orbit, while others travel inward and sometimes cross the orbits of Mars and Earth. *Some, known as* EARTH-GRAZERS, *have at times passed dangerously close to Earth, conjuring up doomsday scenarios.*

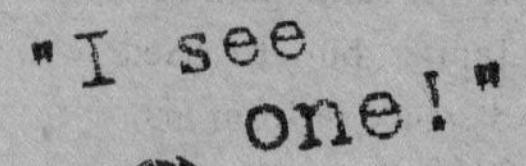

"Hey, it's a flying potato. No, wait, it's Gaspra!"

Gaspra is a bare potato-shaped lump of rock

SHAPELY BODIES

★ *Of the asteroids, only Ceres appears to be spherical*. Judging from images sent back by the Hubble Space Telescope, Vesta isn't far off spherical. They also reveal that it has ancient lava flows, and an immensely deep impact basin, where its mantle is exposed.

★ *Most of the other asteroids have very irregular shapes*. The space probe *Galileo* flew past Gaspra (1993) and Ida (1994) on its way to Jupiter. *Gaspra seems to be the archetypal asteroid, a potato-shaped lump of bare rock about 12 miles long by about 7 miles across, covered with small craters created by meteorite impacts. Ida is nearly three times as big and, surprisingly, has a moon circling around it; only about a mile across, it was given the name Dactyl*.

KEY WORDS

ASTEROID: a tiny body that circles the sun between Mars and Jupiter

EARTH-GRAZER: an asteroid with an orbit that crosses the Earth's

MINOR PLANET: another name for asteroid

THE TROJANS

It was probably the looming presence of giant Jupiter, with its powerful gravity, that prevented another planet from forming in the gap between Jupiter and Mars. In fact, Jupiter's gravity has captured two swarms of asteroids. Called **the Trojans**, they orbit in the same path as Jupiter, but at a fixed distance in front and behind it.

CATCH A FALLING STAR

* You can catch a "falling star" most nights. If you look at the sky long enough, you will see streaks of light that give the impression of stars falling from the sky. But stars don't fall, of course. The streaks are specks of matter from outer space burning up as they hurtle through the Earth's atmosphere.

a falling-star hunter

many UFO sightings can be attributed to fireballs plummeting to Earth

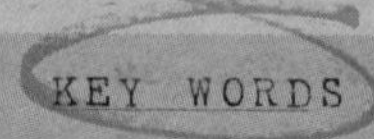

KEY WORDS

METEOR:
a streak of light made by a meteoroid burning up in the atmosphere
METEORITE:
a lump of matter from outer space that has fallen to the ground
METEOROID:
a tiny speck of interplanetary matter

METEOROIDS AND METEORS

* *The space between the planets is full of particles of rock and metal, known as* METEOROIDS. When they come close to the Earth, they are attracted by its gravity and plunge into the atmosphere, where they are slowed down by the air molecules. *Most of them are very tiny, and gently drift to the ground as particles of fine dust, which are called* MICROMETEORITES. Every year some 5 million tons of such dust fall to Earth.

* *Particles larger than a sand grain, however, get heated to incandescence by friction with the air molecules. This creates the fiery streaks of falling stars, or* METEORS, *as they are properly called. Usually, the particles have burned up completely by the time they reach an altitude of 50 miles.*

METEOR SHOWERS

* ***On an average night you can often see half a dozen*** SPORADIC METEORS, ***which may appear in any part of the sky***. But at certain times of the year it's possible to see 50 or more meteors a minute, all appearing to originate from the same part of the sky, known as the RADIANT.

* ***Dense*** METEOR SHOWERS ***like this occur regularly every year and are named after***

spectacular meteor showers often occur in mid-August and mid-November

the constellation in which the radiant is located. The Perseids, with their radiant in Perseus, peak in mid-August and usually put on a good show. So do the Leonids, with the radiant in Leo, in mid-November. ***In good Leonid years thousands of meteors rain down every hour***.

COSMIC BOMBS

Not all meteoroids are tiny. Large ones appear as bright, often vividly colored fireballs, as they hurtle through the atmosphere, trailing a broad tail behind; they are often the source of UFO sightings. Those that survive their fiery ordeal and reach the ground are known as **meteorites**. Big ones gouge out craters. The gigantic 50,000-year-old Meteor Crater, in Arizona, and the much cratered surfaces of the moon and other heavenly bodies, result largely from meteorite impact.

Cometary dust

Meteor showers come from the dust that comets scatter in their wake as they approach the sun. Showers occur when the Earth crosses the path of periodic comets (ones that visit our skies regularly). *The Perseids, for example, have their origins in the dusty trail that Comet Swift-Tuttle leaves behind.*

COMET SPECTACULARS

★ A bright comet passing through our skies is one of the most exciting of all astronomical events. Only a total solar eclipse is more spectacular. But, while an eclipse lasts for a few minutes at most, comets can remain visible for months and grow shining tails that span the heavens.

TELLING TAILS

Typically a comet has a double tail. The most visible component of it—**the dust tail**—is produced by the cloud of dust reflecting sunlight. The paler bluish **ion tail** results from interaction between the gas released by the comet and the stream of charged particles coming from the sun, known as the solar wind. The particles "excite" (ionize) the gas and cause it to glow. Both the radiation and the solar-wind particles exert pressure on the double tail, pushing it away from the comet's head. **This is why a comet's tail always points away from the sun—no matter whether it is traveling toward the sun or away from it.**

EVIL OMENS

★ *Comets have been visiting the Earth's skies since our planet was born*. When people began stargazing, the sudden appearance of a comet unnerved them. It was a portent of evil, of impending disaster. This belief was quite widespread until a few centuries ago. ***Comets were thought to be phenomena of the atmosphere until the Danish astronomer Tycho Brahe showed in the 1500s that they must travel through outer space.***

although comets look big, they are only about 6 miles across

Halley's comet made a fateful appearance at the Battle of Hastings

DIRTY SNOWBALLS

★ ***Comets may look spectacular, but they are just lumps of ice, frozen gases, bits of rock, and dust.*** For most of their lives, these "dirty snowballs"—only tens of miles across—lurk in far-distant regions of the solar system. They only start to become visible when they travel in toward the sun.

★ ***As they do so, heat from the sun begins to melt the ice, transforming it into gas and allowing the dust to disperse. The gas and dust form great billowing clouds that reflect the sunlight, so we see the comet start to shine. And as more gas and dust are released, they build up into a*** TAIL ***that streams out from the comet's*** HEAD.

★ As the comet's orbit carries it farther away from the sun, less and less gas and dust are released. So the comet fades, until it enters deep freeze again and disappears.

PORTENT OF DEFEAT

In 1066, at the Battle of Hastings, the Anglo-Saxon soldiers who fought alongside King Harold had good cause to fear the comet that appeared in the sky. At that time their king was killed and their country conquered by the Normans. We now know that it was one of the returns of Halley's comet.

Hale-Bopp

Readers can't have failed to notice the glorious comet that appeared in our skies in 1997. It was visible for weeks, hanging in the northwestern sky after sunset and in the eastern sky just before sunrise. It was one of the most spectacular comets of the century. Two U.S. astronomers, Alan Hale and Thomas Bopp, had discovered it two years earlier. If you are fortunate enough to be the first person to spot a comet, your name will be preserved for posterity because it will be named after you.

NO HAPPY RETURNS

Some comets never return to our skies. After repeated passes of the sun, they fragment and scatter. Others break up and come to a much more spectacular end. One such was Shoemaker-Levy 9, discovered in 1993 by Californian astronomers Carolyn and Gene Shoemaker and David Levy. This was the ninth comet the team had spotted together. **Soon it became apparent that it was split into more than 20 fragments and was heading for a collision with Jupiter.** Nothing like it had ever been seen before. Here was this line of minicomets heading for the giant planet like cosmic torpedoes. Their aim was deadly. **In 1994, between July 16 and 22, the comet fragments crashed one by one into the Jovian atmosphere, creating spectacular fireballs and leaving dark scars that persisted for days.**

IN FROM THE COLD

★ Most comets appear without warning, like Hyakutake in 1996 and Hale-Bopp in 1997. They follow wide looping orbits, and approach the sun from all directions. Their birthplace is thought to be in a vast cloud of icy bodies on the very edge of the solar system.

THE OORT CLOUD

★ *Billions of cometary bodies reside in a great spherical cloud surrounding the solar system.* This idea was put forward by Dutch astronomer JAN OORT in 1950, and is now universally accepted. No one knows exactly how big the cloud is, but its outer edge could be more than 2 million light-years from the sun—*nearly halfway to the nearest stars*.

"Oh no, it's heading straight for Jupiter!"

TYPES OF COMET

* The Oort Cloud seems to be the source of comets like Hyakutake and Hale-Bopp, which suddenly appear in our skies. ***We call them*** LONG-PERIOD COMETS—

because they won't return to our skies for hundreds, if not thousands, of years. The builders of the Egyptian pyramids and Stonehenge may have seen Hale-Bopp on its last return, some 4,000 years ago.

* ***But other comets are regular visitors to our skies***. These are called SHORT-PERIOD or PERIODIC COMETS, and astronomers write their names with the prefix "P/." Of these, by far the best known is Halley's Comet (P/Comet Halley), which returns to the Earth's skies every 75 to 76 years. Last seen in 1986, it isn't due to return again until 2061. Some comets have much shorter periods. Comet Encke, for example, returns every 3.3 years.

KEY WORDS

Comet:
a small icy body that orbits the sun

Oort Cloud:
a region at the edge of the solar system containing swarms of cometary bodies

Periodic comet:
one that returns to Earth's skies after a relatively short period

HALLEY'S COMET

Halley's comet is not named after its discoverer, but after **Edmond Halley** (1656–1742), England's second Astronomer Royal, **who first realized that it was a regular visitor to the Earth's skies**. After observing it in 1682, Halley accurately predicted that its next return would be in 1758. **Records show that Halley's comet has been observed by astronomers every one of the 30 times it has returned to our skies since 240 BC.**

cosmic bombardment

Read all about it!

COSMIC CATASTROPHES

★ The spectacular impacts of a comet on Jupiter in 1994 reminded us of the havoc wreaked in the past by the debris that occupies the far from empty space between the planets.

Earth hit by asteroid

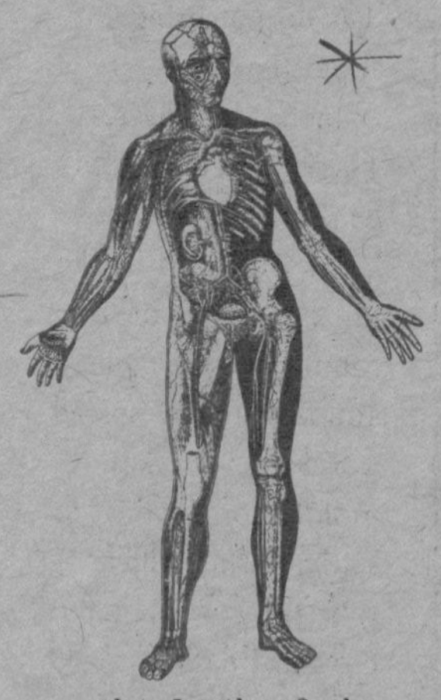

the death of the dinosaurs led to life for mankind

A silver lining

In retrospect, we can see that the mushroom cloud of extinction had a silver lining. The death of the dinosaurs, the "terrible lizards" that had dominated the Earth for over 100 million years, permitted the upsurge of the mammals—and, ultimately, of the king of all mammals: **Homo sapiens**.

RELENTLESS BOMBARDMENT

★ *Since the solar system came into being more than 4 billion years ago, the planets and their moons have been bombarded mercilessly by the chunks of rocky matter—comet remnants and asteroids—that never made the big time*. Our own crater-covered moon bears witness to this. There is not as much debris about today as there once was, but there is more than enough to worry about.

★ The Earth has had its share of bombardment over the eons—though it has fared better than the moon, for example,

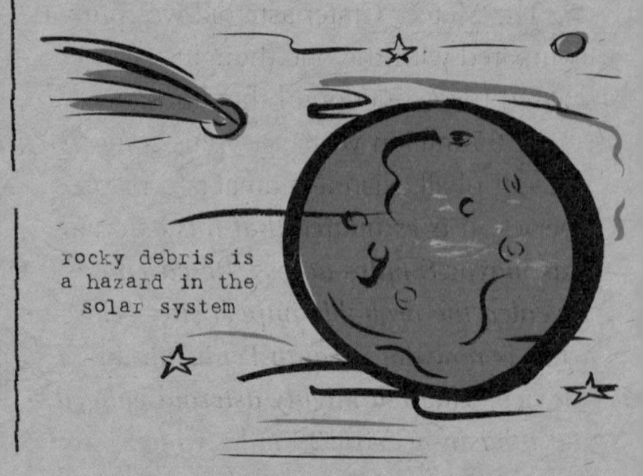

rocky debris is a hazard in the solar system

because its atmosphere has offered some protection. ***Our planet does not reveal its battered past in an obvious manner, because the processes of erosion have long since obliterated ancient craters. But a few that are more recent are still evident—most notably the Meteor Crater in the Arizona Desert.*** Correctly, this should have been called Meteorite Crater—or Asteroid Crater, because it was made by an iron asteroid about 250 feet across.

DINOSAUR DISASTER

* The Meteor Crater asteroid was puny compared with the one thought to have caused the extinction of the dinosaurs about 65 million years ago, plus some 70 percent of all existing animal and plant species. ***It is estimated that this asteroid was maybe 8 miles across. Research has revealed the probable impact site—close to what is now the Yucatan Peninsula in Mexico. There, a mighty asteroid gouged out a basin nearly 125 miles wide.***

MASS EXTINCTION

Imagine the scenario. The blast and tidal waves created by the impact reverberate around the globe. However, although they are devastating, they do not bring about the mass extinction of species. The fatal damage will be done by the dust layers generated by the impact, which mushroom into the high atmosphere and spread worldwide. It takes months for the dust to settle, during which time the Earth remains in virtual darkness. Not only does the temperature plummet dramatically, but photosynthesis, the process by which plants make their food, is no longer possible. The lumbering plant-eating dinosaurs soon starve to death, as do the ferocious carnivores that prey on them.

THE DOOMSDAY ASTEROIDS

* Science-fiction writers are fond of story lines that see Earth attacked by evil empires from elsewhere in the Galaxy. However, it is not bug-eyed aliens that we should fear but the mountain-sized chunks of rock that wander around the solar system.

earth-grazers sometimes come uncomfortably close...

THE TUNGUSKA EVENT

* We should not be lulled into a false sense of security, just because our planet has escaped extraterrestrial disaster in the past. There have, in fact, been a few close calls. ***On the last day of June in 1908, a devastating explosion occurred in the Siberian wilderness, not far from the stony Tunguska River.*** The power unleashed by it was equivalent to a thermonuclear bomb in the tens of megatons range. There was a dazzling fireball, and a blast that felled forests. ***Astronomers reckon it happened when a small asteroid or a fragment of a comet plunged through the atmosphere and exploded at a height of about 5 miles above the Earth's surface. If the event had happened over Moscow—not that far away—the city would have been laid waste and thousands of people killed.***

KEY WORDS

EARTH-GRAZERS: asteroids with an orbit that takes them close to the orbit of Earth (also known as NEOs or near-Earth objects)

the Tunguska explosion was like a huge nuclear blast

EARTH-GRAZERS

* The asteroid believed to have caused the Tunguska incident would have been classed as an EARTH-GRAZER or NEAR-EARTH OBJECT (NEO). *These maverick bodies wander outside the asteroid belt, in orbits that can bring them perilously close.*

which should we fear most, bug-eyed aliens or earth-grazing asteroids?

* This isn't just theory. *On May 21, 1993, astronomer* TOM GEHRELS, *based at the University of Arizona, discovered an NEO some 435,000 miles away*. The good news was that it was receding rapidly. The bad news was more dramatic. *The previous day it had come within 95,000 miles of Earth—by astronomical standards, a near miss*. The reason why it hadn't been spotted earlier was that it had adopted the classic dogfighter tactic. It had come in backed by the glare of the sun.

"The end is nigh"

Gehrels heads a group at the University of Arizona called ***Spacewatch****, devoted to keeping a lookout for threatening NEOs. On December 6, 1997 another member of Spacewatch, Jim Scotti, found one about one mile across, which was cataloged as 1997XF11. After Scotti had calculated the orbit of this body, he felt he had no option but to announce a chilling piece of news.* ***On October 26, 2028 the object cataloged as 1997XF11 would skim past the Earth at a distance of about 30,000 miles. The slightest deviation in its orbit could put it on a collision course with the Earth.*** *The doomsday merchants in the media had a field day running newspaper headlines predicting the end of the world as we know it. Cities in ruins, devastating tidal waves as high as skyscrapers, perpetual night for months . . .* ***But within 24 hours the world breathed a sigh of relief—when new calculations refined the orbit and predicted a miss of about 600,000 miles.***

CHAPTER 8

MYSTERIES OF THE UNIVERSE

*** If, as astronomers believe, solar systems routinely form when stars are born, there must be many other planets scattered around both our own and other galaxies. Could some of them be like Earth?**

"I wonder what **is** up there?"

"There's plenty of evidence that proplyds happen"

KEY WORDS

PROPLYD (PROTO-PLANETARY DISK): a disk of matter in which planets form

DETECTING PROPLYDS

* *It would be nice to know that we could escape to a planet like our own in another solar system if the Earth is ever threatened with extinction by a huge asteroid.* This was the scenario enacted in the 1950s science-fiction movie *When Worlds Collide*. *In practice, this idea is a nonstarter. Even if we had the technology for interstellar travel, we would need to know where to go to reach a planet like Earth—if there is one.*

* *True, astronomers have found plenty of evidence that proplyds (see page 114) form around stars when they are born*. In the 1980s the infrared satellite IRAS detected proplyds around the stars Vega and Beta Pictoris. And the Hubble Space Telescope has photographed such disks in the Orion Nebula and elsewhere. *The matter in these protoplanetary disks would be expected, in time, to condense into planets, as happened with our own solar system.*

NEWLY BORN

★ ***Until recently there was no direct evidence of planets outside our own solar system***. No planets had ever been observed circling around other stars. Not surprising, perhaps, given the odds against a sighting from such a distance. ***But there was some indirect evidence. For example, some stars wobble slightly, which could be caused by circling planets***. But in March 1998 came a dramatic breakthrough. ***The Hubble Space Telescope sent back images of what appears to be a large planet being ejected from a new double-star system***. It is visible because it is still glowing with the heat generated by its formation. It could be as young as 100,000 years old.

PLANETS LIKE EARTH

The probability of there being planets like Earth elsewhere has got to be high. The sun is a very average star, and there are billions like it in our galaxy alone. Assuming that all planets form in a similar way, the solar systems of other stars probably include a mix of inner rocky planets and outer gaseous ones, as in our own Solar System. And in some systems, one of the rocky planets may be Earth-sized and may have formed not too close and not too far away from the parent star. In which case, it could have an atmosphere and climatic conditions similar to the Earth's. **The implication is that there could be many thousands of these Earthlike planets in our own Galaxy—as well as in others. This immediately poses the crucial question. Do these planets teem with life, as Earth does?**

THE MYSTERY OF LIFE

*** It is difficult to speculate about the likelihood of life on the planets of other stars, since we don't really know exactly how life began on Earth. Did it arise spontaneously? Or could it have come from space itself?**

KEY WORDS

SPONTANEOUS GENERATION: the sudden emergence of life

LIFE SPORES: life molecules reaching Earth from space

SPORES OF LIFE

Not everyone accepts the idea of spontaneous generation. One rival theory, advocated by scientists such as Fred Hoyle and Chandra Wickramasinghe, is that **life on Earth originated in "life spores" (primitive organisms like bacteria and viruses) from outer space, which traveled to Earth in meteoroids or cometary debris.** According to their hypothesis, space-borne viruses still bombard the Earth today—and that is why pandemic outbreaks of diseases such as influenza sometimes suddenly occur throughout the world at much the same time, a phenomenon otherwise difficult to explain.

RECIPE FOR LIFE

** There is no doubt that the possibility of life exists throughout the universe. Life as we know it is based on carbon, the atoms of which have the unique ability among the elements to form long chains and create large molecules. Life seems to depend on the complexity of, and interactions between, such molecules.*

"Waiter there's a molecule in my primordial soup!"

* Well, there is plenty of carbon in the universe. It is synthesized in abundance in the interior of massive stars, and blasted into space in their death throes. It is not only found as elemental carbon but also, combined with hydrogen, in simple hydrocarbons. ***Chemically, it is only a short step from these compounds to the more complex life-building molecules***. Oxygen is another atom formed in the interior of stars and blasted out into space. It is found in nebulae, by itself and combined with hydrogen in the form of water. So, throughout space we have hydrocarbon compounds, oxygen, and water—the three main ingredients of life.

HOW LIFE BEGAN

* ***Exactly how did life arise?*** The theory favored at present is "SPONTANEOUS GENERATION" (nothing to do with the old belief that maggots arose from decaying meat, insects from pondwater, and so on).

* The modern concept imagines a young Earth covered in warm oceans and with a heavy atmosphere composed of gases such as nitrogen, methane (a hydrocarbon), and ammonia. Lightning, which illuminates the skies, provides the energy to convert these gases into simple organic chemicals, which rain into the seas, forming a "PRIMORDIAL SOUP." ***Eventually, increasingly complex interactions between the chemicals result in molecules that can replicate themselves. They are the harbingers of life.***

"Add ultraviolet to a cocktail of gases..."

The Miller-Urey experiments

Experiments carried out by Stanley Miller and Harold Urey at the University of Chicago in 1953 provided practical confirmation of the possibility of this scenario. They concocted the cocktail of gases thought to be present in Earth's early atmosphere and subjected them to ultraviolet radiation—to imitate lightning flashes—over steamy water. After only 24 hours, analysis of the resulting mixture revealed the presence of many new carbon compounds, including amino acids. Such acids are the building blocks of proteins, one of the essential components of living tissue.

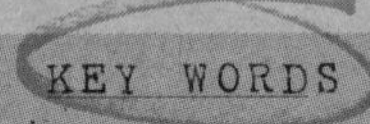

KEY WORDS

EXTRATERRESTRIAL:
from outside the Earth

ET:
a being from another world (abbreviation of extraterrestrial)

SETI:
the search for extra-terrestrial intelligence

UFO:
unidentified flying object—one that can't be readily explained

UFOLOGY:
the study of UFOs (usually implying a belief in extraterrestrial explanations for UFO sightings)

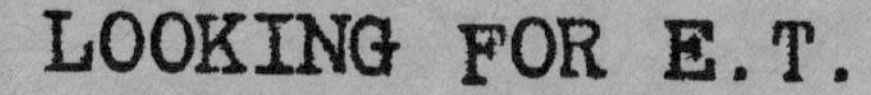

LOOKING FOR E.T.

* Although some people claim to have encountered and even been abducted by aliens—who are often remarkably fluent in terrestrial languages—astronomers have found no evidence that life exists anywhere in the universe other than on Earth. But they are still looking—or rather listening.

"Hello, nice to meet you"

ON THE RADIO

* *Since radio waves are able to penetrate the void of space, it seems likely that any intelligent alien civilization wanting to make contact with us would employ radio broadcasts.* But radio encompasses a broad spectrum of wavelengths, so which one would they choose? *The chances are that it would be the 21-cm wavelength—what we might call the wavelength of the universe, since it is the one emitted by atomic hydrogen, the most abundant element in the universe.* In all probability, our alien broadcasters would use this wavelength as a carrier wave and modulate it with message signals, just as we do in terrestrial radio broadcasting.

LISTENING OUT

radio waves can penetrate the void of space

★ ***It was in 1960 that astronomers first began to listen seriously for alien radio signals***. This was the start of what is now called SETI (the search for extraterrestrial intelligence). Several teams have since been involved in the search, and NASA now has its own SETI program. ***The latest projects use computer-controlled receivers that are able to scan millions of radio frequencies simultaneously. As well as carrying out broad sweeps of the heavens, they target several hundred of the nearest sunlike stars***. No likely signals have been received to date, which some astronomers interpret as meaning there is no one out there.

"No, I hear aliens!"

★ In April 1960 astronomer Frank Drake initiated Project OZMA at Green Bank Radio Observatory, in Virginia. He targeted two nearby sunlike stars—Tau Ceti and Epsilon Eridani—but without success. Nevertheless, Drake's work prompted others to listen for alien signals. ***In 1964 Russian radio astronomers located a possible source, but it turned out to be a fluctuating quasar. In 1967 Cambridge astronomers recorded repetitive signals, but these came from a new type of body never recognized before—a pulsar (see page 96)***.

What about UFOs?

Every now and then, people say they have seen unidentified flying objects, or UFOs, in Earth's skies. Most turn out to be weather balloons, high-flying planes, atmospheric phenomena such as lightning and fireballs, or glimpses of bright planets such as Venus. A few remain unexplained, and some people believe they are reconnaissance craft from alien civilizations. There is no evidence to support the view that UFO sightings have an extraterrestrial explanation. Confirmed ufologists, however, will howl "Cover-up!" to such a statement.

I am disgusted

The fact that we haven't heard back from any listeners or viewers in the cosmos does not necessarily mean that no one is out there. ***Even if a viewer on the planet Splott, 60 light-years away, radios a complaint today about the disgusting quality of the Earth's TV programs, it will be another 60 years before we receive it—120 years after the screening of the programs. And it would take another 60 years to send a reply.*** *Two-way communications across the cosmos are bedevilled by such time delays.*

EARTH CALLING

★ If other alien civilizations do exist in space, maybe they too are listening for signals that will tell them of life on other worlds. And maybe one day they will pick up signals from Earth.

FOLLOWING THE SOAPS

★ ***Earth has been beaming radio signals into the heavens since the 1920s, when broadcasting began***. Today radio and TV transmitters around the globe pour out broadcasts in scores of languages. So our news, views, sports, and soap operas are being disseminated throughout the cosmos.

aliens may have been following our radio dramas

★ ***Radio waves travel at the same speed as light, so by now the first radio broadcasts will have reached out in space to a distance of more than 70 light-years. And, in theory, there should be plenty of stars with habitable planets within that radius.***

★ By now people on planets within a 60-light-year radius should have a (slightly out-of-date) idea of what our own planet looks like, too, since TV broadcasts started in the late 1930s.

THE ARECIBO MESSAGE

★ ***Scientists have also sent out messages to inform ETs about our own civilization.*** In 1974 a message was beamed into space from the huge radio telescope at Arecibo in Puerto Rico. It took the form of graphics expressed in digital code, and included ***a human figure, the double-helix structure of the basis-of-life molecule DNA, and a representation of the solar system.***

ROBOT AMBASSADORS

★ Space probes *Pioneer 10* and *11*, and *Voyager 1* and *2* (now traveling through interstellar space), carry messages for aliens who may encounter them. ***The Pioneers carry a pictorial plaque that shows a man and a woman to scale with the spacecraft and pinpoints their celestial location. The Voyagers carry a record disk, "Sounds of Earth." This contains sights and sounds of the natural and manmade world, together with messages from the Earth's leaders.***

Goodwill to all

For inclusion on the ***"Sounds of Earth"*** *record, the then US president, Jimmy Carter, composed a message that extended a hand of friendship across the stellar frontier. It was, he said,* ***"a present from a small distant world, a token of our sounds, our science, our images, our music, our thoughts, and our feelings."*** *Its purpose was to represent "our hope and our determination, and our goodwill in a vast and awesome universe."*

this new sounds of earth record really swings